P9-BJL-277

KEEPING OPTIONS ALIVE:
The Scientific Basis for Conserving Biodiversity

Walter V. Reid
Kenton R. Miller

W O R L D   R E S O U R C E S   I N S T I T U T E
A Center for Policy Research

October 1989

Library of Congress Cataloging-in-Publication Data

Reid, Walter V. C., 1956–
  Keeping options alive.

  (World Resources Institute Report)
  Includes bibliographical references.
  1. Biological diversity conservation. I. Miller,
Kenton.   II. Title.   III. Series.
QH75.R44   1989   333.95′16   89-22697
ISBN 0-915825-41-4

Kathleen Courrier
Publications Director

Brooks Clapp
Marketing Manager

Hyacinth Billings
Production Manager

Organization of American States; Mike McGahuey, World Bank photo/James Pickerell
Cover Photos

Each World Resources Institute Report represents a timely, scientific treatment of a subject of public concern. WRI takes responsibility for choosing the study topics and guaranteeing its authors and researchers freedom of inquiry. It also solicits and responds to the guidance of advisory panels and expert reviewers. Unless otherwise stated, however, all the interpretation and findings set forth in WRI publications are those of the authors.

Copyright © 1989 World Resources Institute. All rights reserved.
Reprinted 1993

Contents

AUGUSTANA UNIVERSITY COLLEGE
LIBRARY

Foreword

As the 21st Century approaches, the world is being impoverished as its most fundamental capital stock—its species, habitats, and ecosystems—erodes. Not since the Cretaceous Era ended some 65 million years ago have losses been so rapid and great. If the trend continues, one quarter of the world's species may be gone by 2050. Desertification, fisheries collapse, tropical deforestation—such losses already attest amply to how much biological impoverishment costs human beings. If we continue to borrow from the future, literally eating our seed corn, those costs will rise.

An alarm has been sounded. In response, the number of international organizations, national governments, and non-governmental organizations concerned about biodiversity and the breadth of their activities has increased dramatically during the last decade. Under their wings, protected areas, zoos, botanical gardens, aquaria, seed banks, and other sanctuaries and research stations have been set up to rescue and propagate endangered species. Some of the participants in this new movement are institutions working to demonstrate the compatibility of ecosystem protection and economic development; others are integrating development planning and crafting an international convention for preserving biodiversity. Some development assistance agencies are considering how their grants and loans affect biodiversity, and various groups are exploring new ways to finance the conservation of biological diversity.

Still another reason for optimism is the rapid strides now occurring in conservation biology and landscape ecology. Scientists' understanding of how to maintain viable breeding populations of species and provide sufficient habitat to support them has grown dramatically since the late 1970s, as has the availability of effective conservation techniques. Worldwide, more people trained in forestry, ecology, conservation biology, and other key fields are needed, but their ranks are beginning to grow.

For all these reasons, the chances of reversing the current trends in many ways look better now than they did only a few short years ago. But agreement is universal that current efforts are insufficient. What is needed is a coordinated attack on the problem at its roots, one that both makes use of the best modern science and reflects concern for the human well-being of those most affected. Only when a "critical mass" of participants are cooperating in a "critical mass" of initiatives within the framework of a common strategy will biological diversity be saved. This cooperation must involve an active participation among governmental and non-governmental organizations in both developing and industrial countries.

In a way, *Keeping Options Alive* is a "how to" publication. Its timely premise is that the biological sciences can help policy-makers identify the threats to biodiversity, evaluate conservation tools, and come up with successful management strategies to the crisis of biotic impoverishment before it is full-blown. To these ends, Reid and Miller provide an overview of where the world's species and genetic resources are located and why they are valuable, a new analysis of species extinctions in tropical forests that supports previous estimates and reinforces the magnitude of the problem that we face, and a survey of the most recent findings of conservation biology. The authors also suggest how these findings can best be put to work for both *in situ* and *ex situ* conservation, and they add to evidence that the biodiversity crisis is not restricted to tropical forests, but threatens biological resources in temperate zones and marine ecosystems as well. Finally, this report underscores the important interdependence between biological diversity and human cultural diversity and the policy implications of this critical bond.

Ultimately, of course, the solution to the biodiversity crisis will be political. It will require both improving planning and management and redressing the social inequities that force people and nations to use resources unsustainably. Indeed, many of the long-term actions that Reid and Miller call for in *Keeping Options Alive* will not occur in the developing world unless the industrial countries provide their fair share of the financing, technology, and knowledge needed to implement them. But creating a decision-making framework based on the most current and comprehensive scientific understanding of the world's biological wealth is essential when so much of it has yet to be inventoried, much less evaluated, and when some of the potentially most important species and ecosystems have few politically empowered constituents. Only by marrying scientific fact to political and economic reality—as *Keeping Options Alive* strives to help policymakers do—can we hope to maintain the biological wealth on which long-term economic development depends.

WRI would like to express its great appreciation to The Rockefeller Foundation, Inc., John D. and Catherine T. MacArthur Foundation, Town Creek Foundation, Inc., The Pew Charitable Trusts, W. Alton Jones Foundation, Inc., The Moriah Fund, Waste Management, Inc., and HighGain, Inc., which have provided financial support for WRI's efforts in this area.

James Gustave Speth
President
World Resources Institute

Acknowledgments

We are grateful to Mohamed El-Ashry, Nate Flesness, Tom Fox, Vernon Heywood, Dan Janzen, Nels Johnson, Jeffrey McNeely, Gordon Orians, Peter Raven, Ulysses Seal, Dan Simberloff, Michael Soulé, Mark Trexler, and Geerat Vermeij for their valuable comments on all or parts of the draft report. Our thanks to Kathleen Courrier for skillfully editing the report and to Hyacinth Billings, Allyn Massey, and Moira Connelly for their help preparing the text and figures. Special thanks to Dee Boersma, Kai Lee, Jeffrey McNeely, Gordon Orians, and Robert Paine for many invaluable discussions on ecology, conservation, and development.

W.V.R.
K.R.M.

I. Introduction

Our planet is a biologically impoverished image of the world that supported humanity in past generations. We can no longer thrill to the sight of waves of migrating passenger pigeons—extinct since 1914—and no tourist industry will be built around the Caribbean monk seal, the last of which died in 1960. Losses of mangrove habitat in many Southeast Asian nations have diminished the productivity of their coastal fisheries, and the potential to boost agricultural productivity worldwide has been reduced with the disappearance of crop varieties possessing unique genetic adaptations.

Today, we are depleting the world's biological wealth at an ever-increasing rate, and the loss and degradation of the world's biological resources are taking a toll on the well-being of people in both industrialized and developing nations. The world entered the current era of geologic history with biological diversity close to its all time high. But the exponential growth in human population and the even faster growth in consumption of the world's natural resources have led to high rates of loss of genes, species, and habitats. An estimated 25 percent of the world's species present in the mid-1980s may be extinct by the year 2015 or soon thereafter (Raven 1988a, b), and significant losses of genetic diversity of both wild and domestic species are expected over this same period. Species of known and potential future value as foods, medicines, or industrial products are disappearing. Various ecosystem "services"—such as the regulation of water discharge and the absorption and breakdown of pollutants—are being degraded as component species vanish from these ecosystems or as natural habitats are converted to other land uses. The erosion of the genetic diversity of agricultural, forestry, and livestock species diminishes the potential for breeding programs to maintain and enhance productivity.

The diversity of life is an irreplaceable asset to humanity and to the biosphere. It provides both immediate and long-term benefits, and its maintenance is essential to sustainable development worldwide. Those components of life that vanish during the next decades will be gone forever; those that remain will provide future options for humanity. Few question that conserving biological diversity is a valuable undertaking. But *how* valuable? *How* threatened is biodiversity? *How* should society decide which components of life most deserve conservation investments? And *how* can this task be accomplished?

Keeping Options Alive poses these fundamental questions and recommends scientifically informed policies for conserving biological diversity. These policies are derived from current knowledge of biogeography, conservation biology, genetics, and systematics, and from population, community, and landscape ecology. These sciences provide guidance for land-use planning and management and afford insights into the relationship between biodiversity and ecological processes; the nature and scale of the threats to species, genes, and habitats; the basis for establishing conservation priorities; the potential effectiveness of various conservation tools; and related topics of pressing concern. Knowledge in these fields is far from complete, but it is sufficient to guide the quick action needed to conserve the world's biological resources, and that guidance can be refined as more information is obtained.

Keeping Options Alive does not address all aspects of biodiversity conservation. Indeed, such important considerations as the economic valuation of biodiversity, the importance of cultural diversity, methods for financing conservation projects, and the use of conservation strategies are scarcely discussed. However, the policies recommended in this report do reflect the broader social, political, and economic context in which biodiversity is threatened and in which its conservation must take place (McNeely et al. 1989). The complex root causes of the planet's biotic impoverishment are intertwined with the causes of poverty and population growth, and they are linked to the inequitable distribution of resources, land, and wealth. Nations are suffering serious economic losses, individual well-being is declining, and future generations are losing invaluable assets as a result of legal and economic incentive structures that favor unsustainable patterns of resource use and discourage local adaptation to environmental conditions.

Considering that each nation's biological diversity is a critical natural asset—for that country and, in some cases, the world—developing strategies for maintaining, studying, and using biological resources sustainably should be national priorities. For that reason, the principles and guidelines set forth in this report must be applied in conjunction with changes in local, national, and international economic and social policies that address the root causes of biotic impoverishment. In particular, new approaches to conservation financing that don't place undue burdens on the developing world are needed (WRI 1989).

Many of the policy and institutional reforms needed to halt biotic impoverishment would be difficult under any circumstances. With the relentless pressures of population growth and the necessity of meeting people's immediate needs for food and shelter, the challenge is all the more daunting. But biotic impoverishment now is less advanced than it would have been had no action been taken in past decades to stem these losses, and a tremendous untapped potential for further conservation exists. Many genes and species are sure to be lost in the coming decades, but which and how many are matters within humanity's control.

II. Why is Biological Diversity Important?

Biodiversity is the variety of the world's organisms, including their genetic diversity and the assemblages they form. It is the blanket term for the natural biological wealth that undergirds human life and well-being. The breadth of the concept reflects the interrelatedness of genes, species, and ecosystems. Because genes are the components of species, and species are the components of ecosystems, altering the make-up of any level of this hierarchy can change the others. Therefore, whether the goal is to obtain products from individual species, services from ecosystems, or to keep ecosystems in a natural state for future generations, these linkages must be taken into account in management policies. How biodiversity contributes to various products and ecological services and figures in the dynamics of ecological systems and how the biological sciences can help policy-makers set priorities for conserving biological diversity are questions of growing importance as threats to biodiversity mount.

Species are central to the concept of biodiversity. Individual species—Earth's various plants, animals, and microorganisms—provide the rice and fish we eat, the penicillin doctors use to save lives, and other natural products. They also provide options for addressing future human needs, and invaluable aesthetic, spiritual, and educational benefits. Just as important, species provide more subtle benefits in the form of wide-ranging ecological services. Coastal wetland ecosystems formed from various plant and animal species remove pollutants from the water and provide the spawning and rearing habitat for commercially important fish and crustaceans. Similarly, forest ecosystems help regulate water discharge into rivers, which affects the frequency of floods and the availability of water during dry seasons. These and other ecosystems also influence local climatic conditions or, in the case of a forested area as large as Amazonia, even global climate. In a sense, species and ecosystems are integrated service networks and the parts need to be conserved to conserve the whole. Just as habitats and ecosystems must be maintained to conserve species, species must be conserved to maintain habitats and various ecological services.

Each species' characteristics are determined by genetic make-up. Management of this genetic diversity is particularly important in small populations and in domestic species. Humanity has long recognized and utilized genetic diversity in the development of varieties of domesticated plants and animals for use in agriculture, forestry, animal husbandry, and aquaculture. In U.S. agriculture alone, crop-breeding programs drawing on genetic diversity add an estimated $1 billion annually to the value of production (OTA 1987). Another way farmers take advantage of genetic diversity is by planting numerous varieties of crops as a hedge against total crop failure. By planting several varieties of potatoes, for example, Andean farmers can count on a successful harvest almost regardless of what turns the weather takes.

The conservation of biodiversity is *the management of human interactions with the variety of life forms and ecosystems so as to maximize the benefits they provide today and maintain their potential to meet future generation's needs and aspirations.* This definition of conservation, modelled after that used in the World Conservation Strategy (WCS) (IUCN 1980), emphasizes that how *people* use species, manage the land, and invest in development will determine the ultimate success of biodiversity conservation. To many, the word ''conservation'' has a narrower meaning— maintenance or preservation. But ''maintenance'' or ''preservation'' seem most useful when confined to practices that keep an ecosystem or population in its existing state. The broader definition used here, in contrast, entails a variety of objectives. A national biodiversity conservation program, for instance, may involve efforts to remove exotic species harmful to natural or agricultural ecosystems, maintain and utilize the genetic diversity present in crops and their wild relatives, maintain habitats that provide services to humanity and to the biosphere, and save, study, and use the species native to the country.

Seen in this way, the conservation of biodiversity is an important objective for all nations, individually and collectively, and for local communities. The variety of species and genes found in a nation, and the habitats and ecosystems in which they occur, are critical resources that should be utilized sustainably in each country's development. Whether or not a country is species-rich, the management of the human use of the nation's biological diversity should be a national priority to ensure that people's needs are met and that the nation's global responsibility is fulfilled.

The Role of Biodiversity in Ecosystems

Genes, species, and the other components of the world's biodiversity are inseparable from the processes of life that the components give rise to—among them, production, consumption, and evolution. *(See Figure 1.)* Together, biodiversity (that is, the *elements* of life), and ecological processes (the *interactions* among species and between species and their environment) define Earth's living mantle—the biosphere. For individuals and populations, these interactions include such mechanisms as predation, competition, parasitism, and mutualism, while communities change through the process of succession. In yet another type of interaction, species influence their physical environment—whether through primary production (the transformation of solar energy to biomass through photosynthesis), decomposition (the breakdown of organic materials by organisms in the environment), or participation in biogeochemical cycles (the movement of nutrients, water, and other chemical elements through living organisms and the physical environment). Other important ecological processes include soil generation and the maintenance of soil fertility, pest control, climate regulation and weather amelioration, and the removal of pollutants from water, soil, and air.

Ecological Processes

No simple relationship exists between the diversity of an ecosystem and such ecological processes as productivity, water discharge, soil generation, and so forth. For example, species diversity doesn't correlate neatly with biological productivity. Species-rich tropical rain forests are extremely productive, but so are coastal wetlands, which have relatively low species diversity. Species diversity also does not correlate closely with an ecosystem's stability—that is, its resistance to disturbance and speed of recovery. For example, coastal salt marshes and Arctic tundra are dominated by a handful of species, and in some cases—such as *Spartina* salt marshes—one species provides virtually all of the ecosystem's primary productivity (Teal 1962); yet, there is no evidence that these ecosystems are in particular danger of species extinctions or wide population fluctuations in response to disturbances.

4

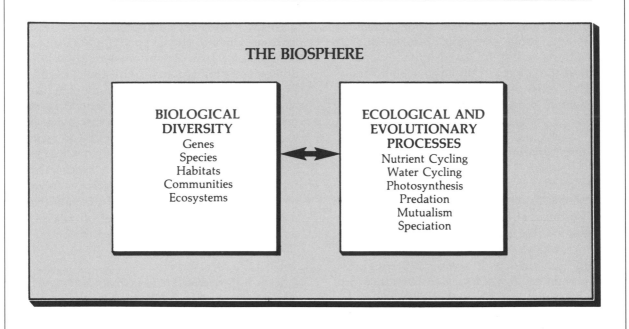

Figure 1: Relationship Between Biological Diversity and Ecological Processes

THE BIOSPHERE

BIOLOGICAL DIVERSITY
Genes
Species
Habitats
Communities
Ecosystems

ECOLOGICAL AND EVOLUTIONARY PROCESSES
Nutrient Cycling
Water Cycling
Photosynthesis
Predation
Mutualism
Speciation

Nor is there a simple relationship within any given ecosystem between a change in its biodiversity and the resulting change in the system's processes. Instead, the outcome depends on which species and ecosystem are involved. For example, the loss of a species from a particular region (known as local extinction or extirpation) may have *little or no effect* on net primary productivity if competitors take its place in the community, as may have happened when eastern hemlock *(Tsuga canadensis)* quickly replaced the dying American chestnut *(Castanea dentata)* in the eastern United States as one of the two dominant species in the forest. In other cases, however, the loss of certain species from an ecosystem could substantially *decrease* primary productivity. If mycorrhizal fungi die out, the growth rate of the plants that they help obtain water and nutrients will decrease dramatically (NRC 1982). Similarly, if such herbivores as zebras *(Equus burchelli)* and wildebeest *(Connochaetes taurinus)* are removed from the African savanna, the ecosystem's net primary productivity decreases (McNaughton 1985). In some cases, the loss of a species could *increase* the ecosystem's productivity if, say, the species normally grazes heavily on the ecosystem's vegetation. For example, if sea urchins, limpets, and other intertidal species are extirpated, algal productivity in intertidal and subtidal zones sometimes increases substantially (Paine 1980).

Altering an ecosystem's species diversity has case-specific effects on such ecosystem processes as water and nutrient cycling too. Often, for example, when a forest ecosystem is simplified, greater amounts of water are lost as runoff in floods. Peak storm runoff on rubber and oil-palm plantations in one small Malaysian watershed was about twice that of an adjacent naturally forested watershed, while low flows were about half of those of the forested catchment (Daniel and Kulasingam 1974). However, exceptions to this pattern exist. One study comparing a species-rich broad-leaf temperate forest to a species-poor pine plantation found that the pine plantation lost less water to runoff due to the increased interception of rain in the pine canopy and thus higher rates of evaporation (Vitousek 1986). Considering this wide range of possible outcomes, land-use decisions cannot be based upon generalities about the effects of the loss of biodiversity, but rather must be made after careful study of the potential impacts of specific land-use changes.

Ecological Dynamics

If the relationship between species diversity and ecological processes defies general rules, ecologists have at least identified many specific relationships that allow them to assess how environmental changes will affect species diversity and how changes in species diversity will affect certain ecological processes. A number of recent advances in ecology that detail such relationships provide decision-makers with an invaluable picture of the mode and tempo of change in ecosystems, and, more important, provide managers with the information needed to wisely manage biodiversity.

First, *regardless of how static they may appear, the mix of species making-up communities and ecosystems changes continually* (Graham 1988). For example, when the Pleistocene Era ended roughly 10,000 years ago, the flora and fauna of what is now Pennsylvania in the United States included pine and hemlock forests and, among many other species, the smokey shrew *(Sorex fumeus)*, ground squirrel *(Spermophilus tridecemlineatus),* and collared lemming *(Dicrostonyx hudsonius)*. After the glaciers retreated, the ranges of these species shifted, but not together. Pine forest moved northwest, while hemlock moved northeast (Jacobson et al. 1987); *S. tridecemlineatus* and *S. fumeus* are still found in or near Pennsylvania, but *D. hudsonius* moved more than 1500 km to the north (Graham 1986). Elsewhere, armadillos *(Dasypus novemcinctus)* and collard lemmings that co-existed during the Pleistocene today live more than 1500 km apart. In the last few thousand

years, the location, composition, and extent of tropical forests have also changed significantly (Hunter et al. 1988).

Clearly, biological communities do not respond to environmental changes in congress. Instead, the species within communities respond in different fashions to environmental changes and the mixture of species comprising a community at any given time reflects these individualistic responses. Given this fact, the objective of biodiversity conservation should not be to maintain the exact composition of communities that exist today, but rather to maintain the species themselves and to allow ecosystem changes to continue. However academic this distinction may appear, it is crucial in the design of policies to respond to such impacts as those of climate change on biodiversity.

Second, *species diversity increases as environmental heterogeneity—or the patchiness of a habitat— does, but though species richness can sometimes be increased by increasing the diversity of habitats within an ecosystem, this intervention can be a double-edged sword. (See Chapter IV.)* Species that thrive in early successional habitats and benefit from disturbance tend to be those least vulnerable to extinction, whereas those that require large tracts of late-successional habitats may be at greatest risk (Foster 1980).

Third, *habitat patchiness influences not only the composition of species in an ecosystem, but also the interactions among species.* This factor may control the dynamics of predator-prey or host-parasite interactions in both natural and agro-ecosystems. In heterogeneous environments, at least some members of prey or host species can escape from their predators and—it follows— extinction. For example, outbreaks of mange in the rodent-like rock hyrax *(Procavia johnstoni)* in the Serengeti may kill the entire population of a single rocky outcrop, but the transmission of the disease between outcrops is too slow to endanger the entire population (Hoeck 1982).

Fourth, *periodic disturbances play an important role in creating the patchy environments that foster* high species richness. Such disturbances may actually appear rare, random, or unimportant from the vantage of human time. But "unusual" events, such as 100-year hurricanes, floods, or major fires may be "regular" if a longer view is taken. Forests that regularly experience fire or treefalls—even at intervals of hundreds of years—may never reach equilibrium. In other words, succession will never reach an endpoint, species composition will never be fixed, and the relative abundance of species will never be constant. Periodic disturbances keep an array of habitat patches in various successional states. The spatial heterogeneity, in turn, contributes to the diversity of species and influences the interactions among species (Sousa 1984). Disturbance caused by hurricanes may play a pivotal role in maintaining the structure of coral reef communities (Connell 1978), and fires are so common in chaparral communities and jack pine forests that many species require fire to complete their life cycle (Biswell 1974, Cayford and McRae 1983). In one Mediterranean shrub ecosystem, the prevention of grazing, fire, and cutting reduced plant-species diversity by 75 percent (Naveh and Whittaker 1979).

Consequently, to maintain species diversity within a region, it is often necessary to allow natural patterns of disturbance to continue, or at least to manage the environment so as to preserve natural patterns of succession. This conclusion has led to a novel tropical forest management technique known as "strip shelterbelt forestry" that attempts to mimic the natural disturbance patterns in tropical forests in order to maintain the maximum species diversity while still allowing timber harvest (Hartshorn et al. 1987, Hartshorn 1989). In the United States, the important role of natural disturbance is recognized in the policy of not fighting natural fires on public lands unless structures or private property are threatened. Periodic fires prevent the build-up of brush and other fuel that would create conditions for more intense and devastating burns. However, following the highly publicized fires in Yellowstone National Park in the fall of 1988, the

U.S. placed a moratorium on this policy, thereby requiring suppression of all fires. This change is clearly misguided from an ecological standpoint and will add significantly to management costs on federal land.

Fifth, *both the size and isolation of habitat patches can influence species richness, as can the extent of the transition zones between habitats.* Often these so-called "ecotones" support species that would not occur in continuous habitats. In temperate zones, ecotones are more species rich than continuous habitats, though the reverse may be true in tropical forests—perhaps because climatic conditions in the ecotones are stressful or because tropical forest species have particularly narrow habitat requirements (Lovejoy et al. 1986).

Sixth, *certain species have disproportionate influences on the characteristics of an ecosystem.* At one extreme are "keystone" species whose loss could transform or undermine the ecological process or fundamentally change the species composition of the community. At the other are "redundant" species whose loss would have little effect on a particular ecological process.

Obviously, management policies should focus on keystone species since changes in these populations disproportionately affect other species in the community. In the worst case, the local extinction of one of these species may cause a "cascade effect" whereby other species within the community dwindle in number or are extirpated themselves. *(See Appendix 1.)* For example, between 1741 and 1911, the sea otter *(Enhydra lutris)* was all but exterminated from the Aleutian Islands by fur traders. In the absence of this dominant coastal predator, the population size of its prey—sea urchins *(Strongylocentrotus polyacanthus)*—increased dramatically, which, in turn, dramatically decreased the abundance of kelp *(Laminaria spp.)*—a major prey of urchins (Estes and Palmisano 1974, Estes et al. 1978). This loss redounded at many levels. Kelp contributes substantially to the coastal ecosystem's primary productivity and provides physical structure for a highly diverse nearshore fish community, and these fish support seals, bald eagles (*Haliaeetus leucocephalus*), and many other populations. The near-extinction of otters in the Aleutians changed the structure of the coastal marine ecosystem dramatically as kelp beds were replaced with high-density aggregations of urchins.

In contrast, other species losses have little effect on the remaining species. Around 1900, a chestnut tree fungus, native to China and Japan, was inadvertently introduced into the United States. The fungus all but wiped out the American chestnut by 1950 (Burnham 1988). Despite the loss of a species that once composed 25 percent or more of the eastern hardwood forest from Mississippi to Maine, the general structure of the forest changed little. After roughly 25 years, the chestnut-oak assemblage was replaced by a hemlock-oak community, ecosystem processes were unchanged, and no extinctions of birds, mammals, or reptiles have been attributed to its loss (McCormick and Platt 1980, Pimm 1986).

The somewhat chaotic view of nature that these six ecological relationships reveal is at odds with the popular conception of a ''balance of nature,'' in which all species are interconnected and ecological processes ensure that ecosystems move on successional trajectories to a steady state. Indeed, ecologists have long understood that the balance of nature is precarious at best. As early as 1930, Charles Elton noted:

''The simile of the clockwork mechanism is only true if we imagine that a large proportion of the cog-wheels have their own mainsprings, which do not unwind at a constant speed. There is also the difficulty that each wheel retains the right to arise and migrate and settle down in another

clock, only to set up further trouble in its new home. Sometimes, a large number of wheels would arise and roll off in company, with no apparent object except to escape as quickly as possible from the uncomfortable confusion in which they had been living.'' (Elton 1930)

With knowledge of the particular roles of species within communities and the important influences of disturbance and environmental heterogeneity on species richness growing, it is increasingly possible to use and manage land in ways that maintain the species within a region and provide valuable ecosystem services to humanity. But recognizing the trade-offs inherent in various changes in an ecosystem's characteristic diversity—the pattern of distribution and abundance of populations, species, and habitats—is essential to achieving sustainable development worldwide. Characteristic diversity can be increased by, for instance, adding exotic species or allowing moderate disturbances. It can be decreased through such changes as species loss or the prevention of natural patterns of disturbance and invasion. An ecosystem's characteristic diversity can be altered to modify the services that the ecosystem provides to humanity. But in the quest to enhance one service, other essential ecosystem services are often compromised. Establishing timber plantations may increase timber productivity, but reducing species diversity in this way may increase the frequency of floods and soil erosion or reduce water flows during dry seasons, and it obviously harms the species that are removed (thereby diminishing the ecosystem's ''genetic library''). Since altering the ecosystem to enhance short-term productivity causes multiple changes in other ecological processes, and since these changes may ultimately reduce long-term productivity, the focus of management policies can't be limited to only a small number of these effects.

III. Where is the World's Biodiversity Located?

As vitally important as the world's rain forests are, they should not be the sole focus of biodiversity conservation. Even in less species-rich ecosystems, the conservation of biodiversity is critical for meeting local people's needs. Moreover, the various foods, medicines, and industrial products that humanity has obtained from the world's biota have come from virtually all ecosystems and taxonomic groups. Policies for conserving biodiversity must be based on a broad understanding of its distribution. Three questions in particular should inform all conservation policy. First, knowledge of the world's biota—and especially of the identity and distribution of species—is far from complete, so how should this knowledge gap influence priorities for conservation? Second, do historical discoveries of economically valuable species provide the basis for predicting where future discoveries will occur? And, third, what special policies are needed to protect species with limited ranges or small populations?

General Patterns of Species Distribution

Some 230 years after Linnaeus began classifying the variety of life on earth, we still do not know how many species exist—even to within a factor of ten. Of the estimated 10 million to 30 million species on earth, only some 1.4 million have been named and at least briefly described (May 1988, Wilson 1988a).

Our incomplete knowledge of the distribution and diversity of species is particularly striking for certain groups of organisms, such as insects. For example, in a recent survey of just 19 trees of the same species in a tropical forest in Panama, fully 80 percent of the more than 950 species of beetles found are believed to be previously unknown to scientists (Erwin and Scott 1980; T. Erwin, Smithsonian, personal communication, Feb. 1989). More detailed studies of tropical insects are now expected to reveal millions of new species, perhaps even tens of millions (Erwin 1982). Moreover, nearly every animal species may be host to at least one specialized parasitic species, only a small fraction of which have been described (May 1988). Even tropical vertebrates are far from fully described. An estimated 40 percent of all freshwater fishes in South America have not yet been classified (NRC 1980).

The deep sea floor is a similarly unstudied region that is proving to be extremely species rich, containing as many as a million undescribed species (Grassle 1989, Grassle et al., in press). Consider the case of hydrothermal marine vents. Discovered in the mid-1970s along ridges on the ocean bottom where earth's crustal plates are spreading apart, these vents are home to life forms that are largely new to science. More than 20 new families or subfamilies from these environments, 50 new genera, and over 100 new species have been described (Childress et al. 1987, Grassle 1989).

Species Richness

For most well-studied groups of organisms, species richness increases from the poles to the equator (Stevens 1989). The species richness of freshwater insects, for example, is some three to six times higher in the tropics than in temperate sites (Stout and Vandermeer 1975). Similarly, tropical regions have the highest richness of mammal species per unit area, and vascular plant species diversity is much richer at lower latitudes. *(See Table 1 and Figure 2.)*

Table 1. Number of Mammal Species in Selected Countries

Country	Mammal Species	Species per 10,000 Km²[a]	Country	Mammal Species	Species per 10,000 Km²[a]
TROPICS					
Tropic of Cancer to Tropic of Capricorn					
Algeria	97	23	Kenya	308	105
Angola	275	76	Lesotho	33	25
Benin	187	98	Liberia	193	102
Bolivia	267	77	Malawi	192	106
Botswana	154	53	Mali	136	38
Brazil	394	66	Mauritania	61	18
Burkina Faso	147	61	Mexico	439	108
Burundi	103	79	Mozambique	183	58
Cameroon	297	107	Namibia	161	50
Cape Verde	9	11	Nicaragua	177	87
Central African			Niger	131	37
Republic	208	69	Nigeria	274	82
Chad	131	36	Panama	217	126
Colombia	358	102	Paraguay	157	59
Congo	198	77	Peru	359	99
Costa Rica	203	131	Rwanda	147	114
Côte d'Ivoire	226	90	Sao Tome	7	14
Cuba	39	20	Senegal	166	75
Djibouti	22	18	Sierra Leone	178	105
Ecuador	280	115	Somalia	173	58
El Salvador	280	230	Sudan	266	61
Equatorial			Suriname	200	95
Guinea	182	138	Tanzania	310	93
Ethiopia	256	75	Tobago	29	74
Gabon	190	79	Togo	196	124
Gambia	108	106	Trinidad	85	103
Ghana	222	96	Uganda	311	134
Guatemala	174	92	Uruguay	77	35
Guinea	188	80	Venezuela	305	92
Guinea Bissau	109	78	Zaire	409	96
Guyana	198	88	Zambia	228	72
Honduras	179	94	Zimbabwe	194	73
Jamaica	29	28	Average (Tropical):		79.5

Table 1. (cont.)

Country	Mammal Species	Species per 10,000 Km²[a]	Country	Mammal Species	Species per 10,000 Km²[a]
TEMPERATE					
Tropic of Cancer to Arctic Circle; Tropic of Capricorn to Antarctic Circle					
Argentina	255	57	Luxembourg	60	87
Australia	299	41	Morocco	108	39
Austria	83	47	Netherlands	60	41
Bulgaria	49	26	New Zealand	69	29
Canada	163	26	Norway	54	21
Chile	90	29	Portugal	56	31
Denmark	49	33	South Africa	279	79
Egypt	105	31	Spain	100	35
Finland	62	24	Swaziland	46	40
France	113	39	Sweden	65	24
Germany, Federal			Switzerland	86	59
Republic	94	40	Tunisia	77	37
Ireland	31	19	United Kingdom	77	33
Italy	97	41	United States	367	60
Japan	186	71	Western Sahara	15	6
Libya	76	19	Average (Temperate):		38.8

a. To allow a comparison of the relative species richness per unit area, the species richness in each country is standardized to an area of 10,000 square kilometers with use of a species-area curve. The slope of the curve (z) was determined from the data in the first column and country area ($z = 0.27$).

Source: WRI/IIED 1988 (number of species in each country).

All of Denmark possesses less than twice as many species as there are tree species in one hectare in Malaysia.

Forty to one hundred species of trees may occur on one hectare of tropical rain forest in Latin America, compared to only ten to thirty on a hectare of forest in eastern North America. One patch of moist tropical forest near Iquitos, Peru, has approximately 300 tree species greater than 10 cm. in diameter per hectare; and a region in lowland Malaysia near Kuala Lumpur has some 570 plant species greater than 2 cm. in diameter per hectare (Gentry 1988, S. Hubbell, Princeton University, personal communication, Feb. 1989). In comparison, all of Denmark possesses less than twice as many species (of all sizes) as there are tree species in one hectare in Malaysia.

Global patterns of species diversity in the marine environment resemble those on land. On Australia's Great Barrier Reef, for example, the number of genera of coral increases from less than ten at the southern end to more than

Figure 2: Global Patterns of Species Richness of Vascular Plants
Each Bar Represents the Number of Species per 10,000 km² in the Country or Region Indicated

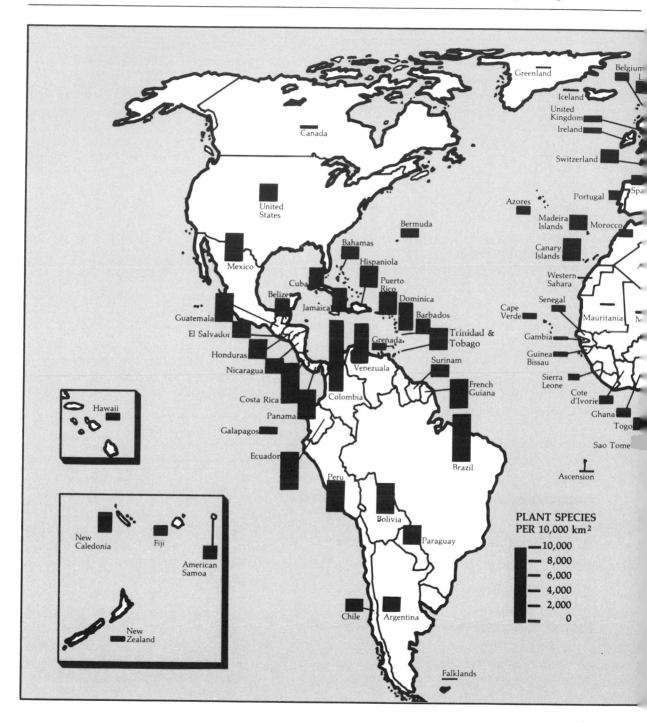

To allow a comparison of the relative species richness per unit area, log (species number) was regressed on log (country area) and the slope of the relationship ($z = 0.33$) was used to adjust the number of species in each country to a standard area of 10,000 km².

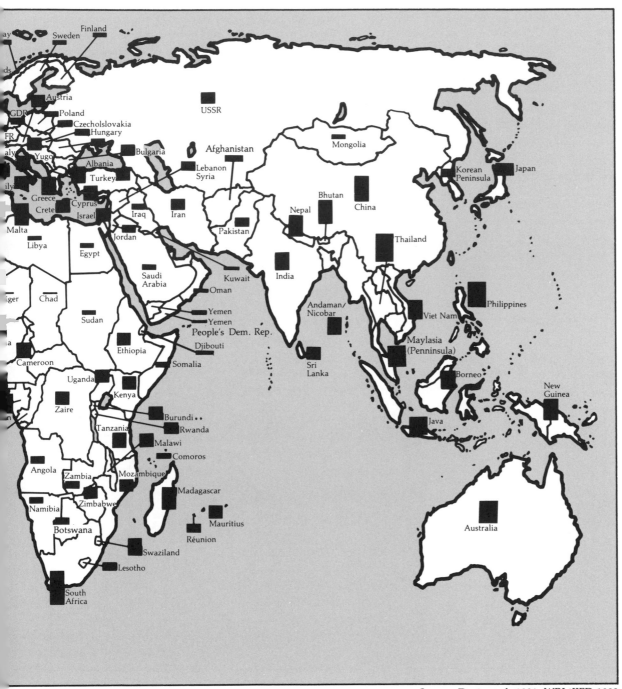

Source: Davis et al. 1986, WRI/IIED 1988.

fifty at the northern end (Stehli and Wells 1971). *(See Figure 3.)* Similarly, the number of tunicate (sea squirt) species increases from 103 in the Arctic to some 629 in the tropics, the diversity of planktonic foraminifera (a group of marine micro-organisms) increases from only two species near the poles to some sixteen in tropical waters (Fischer 1960, Brown and Gibson 1983), and deep sea species diversity also tends to be higher at lower latitudes (Grassle 1989).

These terrestrial and marine patterns of increasing diversity in the tropics reach their peak in tropical forests—the most species-rich ecosystems in the world *(Box 1)*—and coral reefs *(Box 2)*.

Locally and regionally, the general pattern of increasing diversity at lower latitudes does vary. For example, South Africa, well south of the tropics, has the highest plant species richness per unit area outside of the neotropics. Moreover, for both plants *(Figure 2)* and mammals *(Table 1)*, tropical Africa tends to be less species-rich than tropical America or tropical Asia. Tropical marine species richness is highest in the tropical Indo-West-Pacific and decreases to the east, with the lowest richness found in the Eastern Atlantic. *(See Figure 3 and Table 2.)* For many taxa, species richness in the Indo-West-Pacific is more than ten times greater than in the Eastern Atlantic. The Philippines, Indonesia, New Guinea, and the Solomon Islands are particularly species-rich (Vermeij 1978).

Figure 3: Global Pattern of Distribution of Corals
Contours of Number of Genera of Reef-forming (Hermatypic) Corals

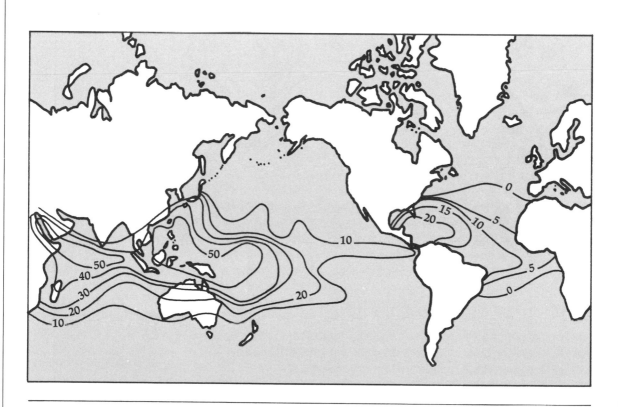

Source: Stehli and Wells 1971.

14

Box 1. Tropical Forest Species Richness

Closed tropical forests contain more than half of the world's species, though they cover only 7 percent of the earth's land surface. The relative species richness of the tropical forest biome varies with the group of species involved, and for some taxa, scientific knowledge of species richness is limited.

The most complete information available is on plant species. The neotropics contain an estimated 86,000 species of vascular plants (Gentry 1982), tropical and semi-arid Africa contains 30,000 species (Brenan 1978), Madagascar contains 8,200, and tropical Asia, including New Guinea and tropical Australia, account for 45,000 species (P. Raven, Missouri Botanical Garden, personal communication, April 1989). In all, tropical regions support two thirds of the world's approximately 250,000 species of vascular plants. Myers (1980) estimated that two thirds of all tropical plant species are found in moist tropical forests (evergreen or deciduous closed forest), and data provided by Gentry (1982: *Table 7*) support this estimate. Accordingly, about 45 percent of the world's vascular plant species occur in closed tropical forests.

The percentage of terrestrial vertebrates found in tropical forests compares with that of plants. An estimated 2,600 avian species—1,300 species in the neotropics, 400 in the afrotropics, and 900 in tropical Asia—depend on tropical forests (A. Diamond 1985). This count amounts to roughly 30 percent of the estimated global total. The percentage is

lower than that for plants, but it does not include avian species that occur in the tropical forest but are not completely dependent upon it. Beehler (1985) notes that fully 78 percent of New Guinea's nonmarine birds occur in rain forest, though many of these may survive in other habitats as well.

Among invertebrates, substantial uncertainty exists over the relative abundance of species in tropical forests. Until recently, the relative diversity of arthropods in the tropics as compared to the temperate zone was expected to be similar to that of better known groups, such as vascular plants or birds. However, Erwin's (1982) discovery of a tremendous richness of beetle species in the canopy of a moist tropical forest suggests that the relative richness of arthropods in the tropics is much greater. As many as 30 million arthropod species—up to 96 percent of the world's total for *all* species—may exist in tropical forests (Erwin 1982).

The fraction of the entire world's species that occur in tropical forests can't be accurately estimated because the total number of species in several potentially large taxonomic and ecological groups—including insects, nematodes, and bottom-dwelling marine invertebrates—is unknown (May 1988, Grassle et al., in press). However, half of all vertebrates and vascular plant species occur in tropical forests and if the tremendous species richness of arthropods in this biome is any indication, at least 50 percent—and possibly as much as 90 percent—of the world's total species are found in closed tropical forests.

These regional and local patterns of diversity are created by historical factors (Raven and Axelrod 1974) and by gradations of precipitation, elevation, and ocean depth (Diamond 1988a). On land, species richness tends to

decrease with elevation and with precipitation. The Sahel, for instance, has fewer species of plants per unit area than surrounding regions do. In the sea, the species richness of bottom-dwelling organisms increases with depth to

Box 2. Coral Reef Species Richness

With extremely complex physical structures, high levels of primary productivity, and the highest species diversity of any biome in their respective environments, coral reefs are in many ways the marine analog of tropical forests. The net primary productivity of coral reefs is approximately 2,500 grams of carbon per square meter per year (gm/m²/yr), compared to 2,200 gm/m²/yr for tropical forests and only 125 gm/m²/yr in the open ocean (Whittaker 1975).

The species richness of coral reefs is unparalleled in the marine environment. The Great Barrier Reef, the world's largest system of coral reefs (covering 349,000 square kilometers), supports more than 300 species of coral, 1,500 species of fish, and over 4,000 species of mollusks (IUCN/UNEP 1988). In addition, 252 species of birds nest and breed on the coral cays, five species of turtles live on the reef, and several species of whales and dolphins are associated with it. Coral diversity is greatest in the Indo-West-Pacific, and species associated with coral follow similar trends in diversity. (See Figure 3.) In the Philippines, more than 2,000 fish species live on or near coral reefs, compared to only 448 in the waters surrounding Hawaii and 507 in the Bahamas (Goldman and Talbot 1976).

An extremely high diversity of fishes—at least 3,000 species—are found among the coral reefs of the Indo-West-Pacific (Goldman and Talbot 1976). Indeed, the Indo-West-Pacific supports more than 16 percent of the world's estimated 19,000 species of freshwater and marine fish. The Great Barrier Reef alone, occupying only one-tenth of one percent of the ocean surface, supports nearly 8 percent of the world's fish species.

In comparison, the coastal waters of the Mediterranean sea support less than 25 percent as many fish species as the Great Barrier Reef and less than 20 percent as many as the Philippines (Briggs 1974). Similarly, the mid-Atlantic seaboard of the United States, roughly comparable in length to the Great Barrier Reef, has only 250 species of fish—less than one fifth as many as the Great Barrier Reef supports (Briggs 1974).

The diversity of species in smaller portions of coral reefs is equally impressive. The Capricorn reefs at the southern end of the Great Barrier account for only 3 percent of the area of the Great Barrier Reef complex yet support 859 species of fish and 72 percent of the complex's coral species (Goldman and Talbot 1976, IUCN/UNEP 1988). This richness of fish species (4.5 percent of the world's total) compares roughly with Costa Rica's richness of plant (3 percent of the world's total) and mammal (4.7 percent) species, yet Costa Rica is four times as large as the Capricorn portion of the Great Barrier Reef.

Although coral reefs share numerous attributes with tropical forests, the level of local species endemism is much lower on reefs. Within the Indo-Pacific, for example, the vast majority of coral species are found throughout the region (Vernon 1986). Because coral reef species disperse readily, locally endemic species occur only on isolated oceanic islands. For instance, 20 percent of the corals and 30 percent of the inshore fishes in Hawaii are endemic to that island chain (Jokiel 1987, Hourigan and Reese 1987). Because coral-related species tend to be widely distributed, they are less threatened than tropical forests are by species extinction. However, degradation threatens both ecosystems' ability to meet human needs.

Coral reefs stand out from other marine environments because of their species

Box 2. (cont.)

diversity, but many coral reef species also depend on other affiliated ecosystems. Often, coral reefs, mangroves, and seagrass beds are linked physically and biologically. Reefs serve as breakwaters that allow coastal mangroves to develop; the calcium of the reef provides the sand and sediment in which mangroves and seagrasses grow; and the mangroves and seagrass communities provide energy input into the coastal ecosystem and serve as spawning, rearing, and foraging habitat for many of the species associated with the reefs (Johannes and Hatcher 1986).

---

roughly 2,000 to 4,000 meters on the continental slope and decreases thereafter (Vermeij 1978, Brown and Gibson 1983).

From a conservation standpoint, one particularly notable feature of species distributions is that regions rich in some groups of species aren't necessarily rich in others. If species per unit area is the benchmark, Central America is more species rich in mammals than northern South America, but northern South America has more plant species. *(See Figure 2 and Table 1.)* In Africa, the species richness of butterflies is greatest in West Africa just north of the equator, while the diversity of passerine birds, primates, and ungulates is greatest in Central and East Africa, and plant diversity is greatest just north and south of the equator in West Africa (IUCN/UNEP 1986b).

Tropical species diversity, in both marine and terrestrial environments, is the world's highest. But because species diversity varies considerably on smaller scales and patterns of species richness do not necessarily coincide among taxa, conservation actions taken on the basis of current knowledge of species diversity will be "correct" only for plants, vertebrates, and other comparatively well-studied groups. Accordingly, the pressing need now is to acquire information on the many other groups of species.

Commonness or Rarity

Not surprisingly, rare species are more prone to extinction than common ones. But what exactly is "rarity"? A species found in a restricted region can be considered rare even

---

Table 2. Species Richness in Tropical Waters

| Group | Number of Species | | | |
	Indo-West Pacific	Eastern Pacific	Western Atlantic	Eastern Atlantic
Mollusks	6,000+	2,100	1,200	500
Crustaceans				
Stomatopods	150+	40	60	10
Brachyura	700+	390	385	200
Fishes	1,500	650	900	280

Source: Vermeij 1978.

though its population may be large where it occurs. The silver sword (*Argyroxyphium macro-cephalum*) grows only in the crater of Haleakala volcano on Maui, but some 47,000 individuals occur at the site (Rabinowitz et al. 1986). Also rare is a sparsely distributed species, even though it may have a fairly large geographic range. Tigers (*Panthera tigris*), cougars (*Felis concolor*), and other large predators may have historically occurred over large regions, but nowhere were they abundant.

Locally endemic species (those found only in a restricted area) are particularly susceptible to extinction when their limited habitat is disturbed or lost. Wherever environmental conditions have contributed to high rates of speciation or the biota has been isolated for long periods, many species of plants and animals have evolved that are found nowhere else. Thus, locally endemic species often occur on mountains, islands, peninsulas and in other areas where dispersal may be restricted by geography, or where unique local conditions (such as serpentine soils) lead to the evolution of species suited to that specific environment. Regions with Mediterranean climates, for instance, have a high percentage of locally endemic plant species.

Certain islands have even higher percentages of locally endemic species than Mediterranean zones do. Remote oceanic islands—such as Hawaii and Ascension—have the world's most distinctive floras: only a small percentage of the native species on these islands are found anywhere else. (*Figure 4; see also Table 10.*) Extremely remote islands, however, tend to have fewer species than less remote islands of the same size. For example, 91 percent of the 956 flowering plants native to the Hawaiian Islands are endemic to the islands, whereas Crete—only one-half the area of the Hawaiian Islands—has roughly 1700 native vascular plant species, though only 9 percent are locally endemic (Davis et al. 1986, Wagner et al., in press).

Islands with both high species richness and highly distinctive floras are among the most critically important sites for conserving biodiversity. Approximately 80 percent of the nearly 8,000 vascular plant species of Madagascar are found nowhere else, 90 percent of the approximately 9,000 flowering plants of New Guinea are endemic to that island, and 76 percent of New Caledonia's 3,250 vascular plants exist only on New Caledonia (Davis et al. 1986).

Less is known about patterns of species distribution in continental regions—particularly the tropics—than on islands. Continental regions have been surveyed less completely, so many species distributions cannot be accurately mapped. And though a species' presence in a given country may be known, its precise range may not be. Thus, while Zaire has a higher percentage of endemic plants than Côte d'Ivoire, Zaire is so much bigger that this comparison is somewhat misleading (*See Figure 4*). The range of an average species in Cote d'Ivoire may be identical to that of a species in Zaire, but a smaller percentage of species are endemic to the country simply because a smaller country contains fewer ranges.

Roughly one percent of West Germany's species are locally endemic, compared to 15 percent in Costa Rica—a country only half West Germany's size.

This accounting problem notwithstanding, evidence is growing that tropical species have more localized distributions than temperate species (Rapoport 1982, Stevens 1989). As Figure 4 shows, aside from South Africa, the African countries with the highest percentage of locally endemic plants are tropical. Similarly, an estimated 15 percent of the floras of several Central America countries are endemic to those countries—a much greater percentage than is found in similarly-sized temperate countries.

Roughly one percent of West Germany's species are locally endemic, compared to 15 percent in Costa Rica—a country only half West Germany's size. One comparison of several well-studied sites in both Latin America and the north temperate zone revealed that the fraction of the flora in the tropical sites restricted to areas of less than 50,000 square kilometers, equals or exceeds the fraction in all the temperate sites except the Cape Region of South Africa (Gentry 1986). Some 440 of South America's land bird species—roughly 15 percent its avifauna—occupy ranges of less than 50,000 square kilometers while the United States contains only eight species with similarly restricted ranges—one percent of the region's avifauna (Terborgh 1974).

Even in apparently continuous tracts of tropical forest, regions with many locally endemic species have been found. How these "centers of endemism" came into being is still debated. One hypothesis holds that the pattern reflects regional differences in climatic and soil conditions and continual fragmentation caused by changing river courses, fires, and flooding (Endler 1982, Gentry 1986, Colinvaux 1987, Räsänen et al. 1987). An alternative theory is that the centers of endemism resulted from fragmentation of the Amazonian rain forest during the last ice age, when a colder and drier climate may have left only small habitat islands of moist forest, particularly at higher elevations. According to this "refugia theory," locally endemic biotas developed in these remaining patches of forest (Haffer 1969, Prance 1982). The approximate concordance of centers of endemism of a variety of taxa in the Amazon basin gives credence to the theory (Simpson and Haffer 1978, but see Beven et al. 1984), but recent studies indicate that Amazonia's climate was probably not drier during the Pleistocene Era (Colinvaux 1987, 1989)—a finding that supports the first hypothesis.

Just as patterns of species richness within a region do not always correspond among different groups of organisms, neither do patterns of endemism. In the northern Andes, most

canopy trees and lianas are widespread whereas many of the epiphytes, shrubs, and herbs are local endemics (Gentry 1986). Plants tend to have more restricted ranges than vertebrates because soil and moisture conditions that only indirectly affect vertebrates often profoundly influence plant distribution and because plants' immobility limits dispersal. In South Africa's *fynbos* (Mediterranean-climate shrubland), roughly 70 percent of the highly diverse plant species are locally endemic, whereas few animal groups of South Africa are restricted to that zone and some are not very diverse (Brown and Gibson 1983, Davis et al., 1986).

Compared to terrestrial species, marine organisms tend to be more widely distributed because they encounter fewer physical barriers. They are thus also less endangered. Fish or such free-floating marine organisms as plankton can travel large distances, often with the aid of currents, and most rooted or stationary organisms have readily dispersed larvae. Consequently, most marine organisms are "cosmopolitan" at the family level and many are at the genus level. Indeed, many marine *species* are found throughout the tropics, including some marine snails, crabs, sharks, and fish (Vermeij 1978). Most species on Australia's Great Barrier Reef are found throughout the Indo-West-Pacific (Vermeij 1978).

Defying this general rule, marine faunas with locally endemic species have developed in certain regions and in certain ecological zones. Nearly half of the species of snails in the high-intertidal zone along Kenya's coast are restricted to the Indian Ocean, and a similar pattern is found among snails in the Red Sea and on the coast of Brazil (Vermeij 1978). Similarly, much of the fauna of the mainland coast of Queensland is found only in that region (Vermeij 1978).

Marine ecosystems with the highest percentage of locally endemic species are found where physical barriers to dispersal exist. The relatively shallow sill of the western Mediterranean, for

Figure 4: Global Patterns of Species Endemism of Vascular Plants
Each Bar Represents the Percent of Species Endemic to the Country or Region Indicated

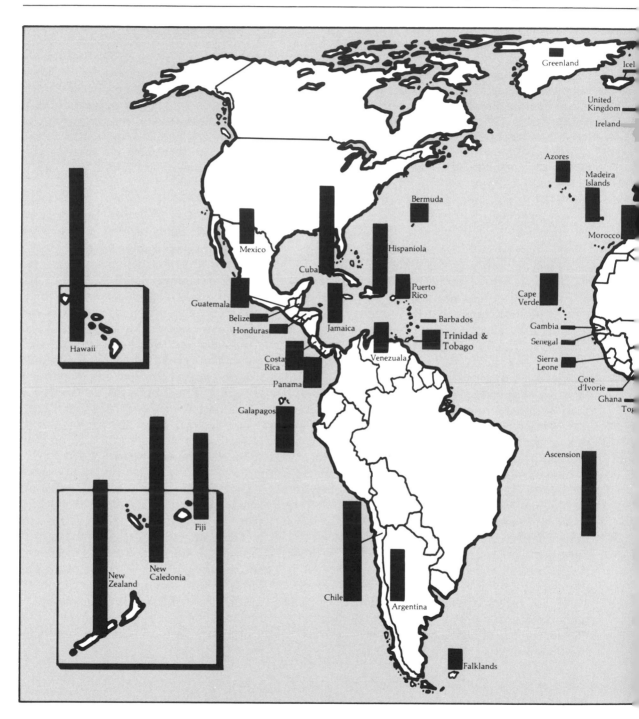

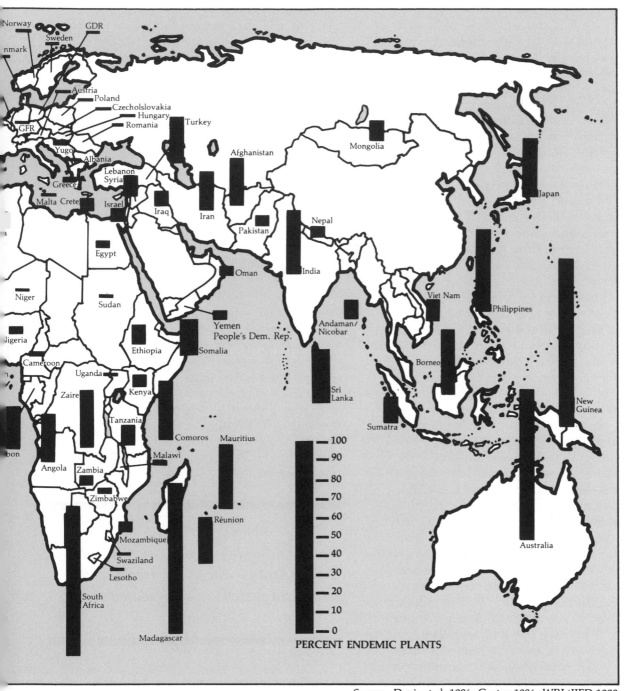

Norway
GDR
Sweden
nmark
Austria
Poland
Czechoslovakia
Hungary
Romania
GFR
Yugo
Albania
Turkey
Afghanistan
Lebanon
Syria
Greece
Mongolia
Malta Crete
Israel
Iraq
Iran
Japan
Egypt
Pakistan
Nepal
Oman
India
Niger
Sudan
Viet Nam
Nigeria
Yemen
People's Dem. Rep.
Andaman/
Nicobar
Philippines
Cameroon
Ethiopia
Somalia
Uganda
Borneo
Zaire
Kenya
New
Guinea
Tanzania
Comoros
Mauritius
Sri
Lanka
bon
Angola
Zambia
Malawi
Sumatra
Zimbabwe
Réunion
Mozambique
Swaziland
Lesotho
Australia
South
Africa
Madagascar

— 100
— 90
— 80
— 70
— 60
— 50
— 40
— 30
— 20
— 10
— 0

PERCENT ENDEMIC PLANTS

Source: Davis et al. 1986, Gentry 1986, WRI/IIED 1988.

example, has partially blocked dispersal and allowed the evolution of numerous species endemic to that sea. In a variety of invertebrate taxa in the Mediterranean, between 13 and 50 percent of the species are locally endemic. Similarly, some 14 percent of the 362 species of fish associated with Mediterranean shores are found nowhere else (Briggs 1974). Fifteen percent of fish species in the Red Sea and 17 percent of fish in the Gulf of California are considered endemic to those bodies of water (Briggs 1974). Isolated oceanic islands and undersea mountains also have many locally endemic marine species. Some 30 to 40 percent of the fish species at Easter Island are locally endemic. In the waters surrounding Hawaii, between 20 and 45 percent of a variety of groups of invertebrates and fish are found nowhere else, and some 41 percent of the stony corals surrounding the Galápagos are endemic to those islands (Briggs 1974).

Besides patterns of endemism, the second aspect of rarity, the abundance or scarcity of a species is also important from the standpoint of biodiversity conservation. In contrast to local endemics, the primary threat to sparsely distributed species is generally not the loss of all suitable habitat, but rather habitat fragmentation, which may so reduce the breeding population within any fragment that the population can't survive. *(See Chapter IV.)* Among animals, species near the end of the food chain—such as large cats, bears, sharks, and eagles—tend to be relatively scarce. But generalizations about plants are harder to make: in any region, some plant species may occur at quite low population densities while others may have large populations (Rabinowitz et al. 1986). In regions rich in plant species, the number of individuals of any given species is often quite small (Hubbell 1979, Hubbell and Foster 1986). In one forested area in Panama, one third of the tree species account for less than one percent of the total number of trees (Hubbell and Foster 1986). Such sparsely distributed species are extremely susceptible to local extirpation or extinction caused by habitat fragmentation.

Strategies for maintaining biological diversity must pay particularly close attention to rare species because of their susceptibility to extinction. Indeed, the commonness or rarity of species is often just as important as species richness as a guideline for biodiversity conservation. But because rarity can result from either restricted distributions or sparse populations, no single strategy will do the trick. Often, species with very restricted distributions can be maintained by protecting the small area where they occur, but maintaining species with sparse populations may require protecting relatively large areas. As with species richness, however, the scientific knowledge regarding species ranges is far from complete, so efforts to increase knowledge of species identity and distribution must accompany conservation actions.

Species of Current Economic Value

Many of the world's most economically important species are found in areas where species diversity is not especially great. None of the world's major food crops originated in tropical rain forests, largely because these species-rich regions were not traditional centers of human population when crops were first domesticated. For similar reasons, only two major crops grown in the United States—the sunflower and the Jerusalem artichoke—originated there.

Only two major crops grown in the United States—the sunflower and the Jerusalem artichoke—originated there.

Understanding where species of current economic value originated and how they diversified is important for several reasons. First, regions possessing wild relatives of domesticated species and regions containing many varieties of a crop should be conservation priorities. If this supply of genetic materials is lost, breeding can't be used to enhance agricultural productivity. Second, knowledge of where valuable species originated can point us to likely locations for future discoveries of species of direct economic value. No attempt is made here to examine all of the current or potential future uses of species. (See Myers 1979, 1983, Oldfield 1984, Prescott-Allen and Prescott-Allen 1986, and OTA 1987.) Instead, the emphasis is on *patterns* of distributions of such species—one basis of decisions on conservation priorities.

Food

Human beings have used about five thousand species of plants as food, but only 150 or so have entered world commerce and less than twenty provide most of the world's food (Frankel and Soulé 1981, Wilkes 1983). Just three crops—wheat, rice, and maize—account for roughly 60 percent of the calories and 56 percent of the protein that humans consume directly from plants (Wilkes 1985). Many of the most important food crops belong to just a few plant families. The grass family—including such crops as wheat, rice, maize, barley, sorghum, millet, oats, and rye—provides some 80 percent of calories consumed by humans, and the legume family has yielded soybeans, peanuts, common beans, peas, chickpeas, cowpeas, and other protein-rich crops. Forty percent of an estimated 2,300 species of cultivated plants belong to just four families: Graminae (grasses), Leguminosae (legumes), Rosaceae (apples, pears, etc.), and Solanaceae (potatoes) (Arora 1985). The remaining species belong to a diverse array of more than 160 families.

Like economically valuable species in general, many of the major food crops originated in regions that are not particularly species rich. Crops were domesticated in warm temperate and subtropical zones and in tropical mountainous regions. Wheat and barley were first grown in the steppes and woodlands of south western Asia (Hawkes 1983), and the origin of maize has been traced to the seasonally dry central highlands of Mexico (Wilkes 1979). The highlands of Peru contributed the tomato and potato, though the tomato was probably first cultivated in Mexico (Hawkes 1983, Wilkes 1979).

Most important food crops appear to have originated where seasons are pronounced, so it makes sense to look there—and not in rain forests—for promising new crops. The cereals of both the Old and New Worlds come from regions with well-marked wet and dry seasons, while root and tuber agriculture, a mainstay of tropical regions, seems to have developed in tropical lowlands with distinct dry periods (Hawkes 1983). As for why, scientists point in part to the tendency of plants in seasonal environments to store nutrients during the growing season; often, these reserves are what human beings eat.

Several regions—known as Vavilov Centers of Diversity after N.I. Vavilov, the Russian botanist who first described the pattern—have been identified as locations of highly diverse crop genetic resources. *(See Figure 5.)* The centers of crop genetic diversity—including the Mediterranean, the Mexican Highlands, Central China, and the Northern Andes—are characterized by a long agricultural history, ecological diversity, mountainous terrain, cultural diversity, and a lack of heavy forest cover (Harlan 1975, Wood 1988). These centers may or may not be located where the crop was first domesticated: wheat and barley were domesticated in southwest Asia, but a current center of their varietal diversity is in Ethiopia (Wood 1988); the tomato originated in northwest Peru, but the greatest domestic varietal diversity is in Mexico (Isaac 1970).

Much of the world's agriculture is based on introduced crops. In developing countries in the Americas, only 32 percent of production, by value, is of crops of American origin (Wood 1988). The comparable figure for African developing countries is 30 percent. Only in

Figure 5: Vavilov Centers of Crop Genetic Diversity
Shaded Areas Indicate Regions of High Current Diversity of Crop Varieties

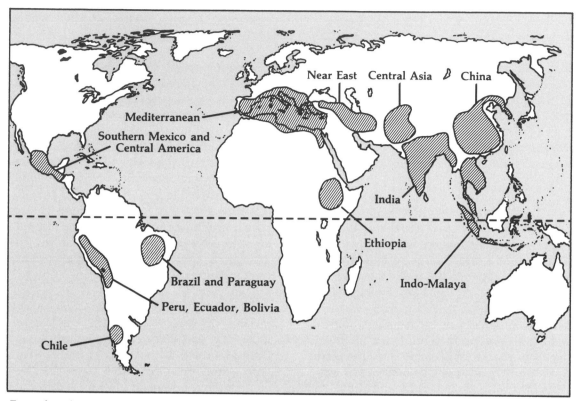

Examples of crops with high diversity in each area include:

China: Naked oat, soybean, adzuki bean, common bean, leaf mustard, apricot, peach, orange, sesame, China tea.

India: Rice, African millet, chickpea, mothbean, rice bean, horse gram, asparagus bean, eggplant, rat's tail radish, taro yam, cucumber, tree cotton, pepper, jute, indigo.

Indo-Malaya: Yam, pomelo, banana, coconut.

Central Asia: Wheat (bread, club, shot), rye, pea, lentil, chickpea, sesame, flax, safflower, carrot, radish, pear, apple, walnut.

Near East: Wheat (einkorn, durum, Poulard, bread), barley, rye, red oat, chick pea, pea, lentil, blue alfalfa, sesame, flax, melon, almond, fig, pomegranate, grape, apricot, pistachio.

Mediterranean: Durum wheat, hulled oats, broad bean, cabbage, olive, lettuce.

Ethiopia: Wheat (durum, Poulard, Emmer), barley, chickpea, lentil, pea, teff, African millet, flax, sesame, castor bean, coffee.

Southern Mexico and Central America: Corn, common bean, pepper, upland cotton, sisal hemp, squash, pumpkin, gourd.

Peru, Ecuador, Bolivia: Sweet potato, potato, lima bean, tomato, sea island cotton, papaya, tobacco.

Chile: Potato

Brazil and Paraguay: Cassava (manioc), peanut, cacao, rubber tree, pineapple, purple granadilla.

Source: Adapted from *Hawkes* 1983

Asian developing countries is most production—70 percent—by native species. Dependence on introduced species reaches its extreme in Australia, the Mediterranean, northern Europe, northern Asia, and the United States and Canada. In these regions, more than 90 percent of production is derived from introduced species. None of the world's twenty most important food crops are native to Australia or to North America north of Mexico (Kloppenburg and Kleinman 1987).

None of the world's twenty most important food crops are native to Australia or to North America north of Mexico.

Whether introduced or native, the most important crops in any region of the world originated or diversified in places with climates similar to those where they are now grown. The main crops grown in temperate zones are thus not the same as those of the tropics where most developing countries lay. Rice, which has origins in either India or China, is the eighth most important crop in the developed world (by weight) but by far—a factor of two—the leading crop in the developing world, and it is the most important source of calories

in tropical developing countries (Isaac 1970, Hawkes 1983, FAO 1987a). Similarly, cassava (also known as manioc)—native to tropical America—is not grown in developed countries but is the fourth most important crop in the developing world (by weight) and provides more than half of the caloric requirements for over 420 million people in twenty-six tropical countries (Cock 1982, Gulick et al. 1983). In Africa, cassava is a fundamental subsistence crop, and in tropical developing countries it is the fourth most important dietary source of calories after rice, maize, and sugarcane.

The fraction of the human diet made up of animals is less than that accounted for by plants. Animals provide one third of the protein in the human diet—roughly 20 percent in developing countries and nearly 55 percent in the developed world. *(See Table 3.)* Most animal foods are obtained from just a few domesticated animals—primarily camels, cattle, chickens, ducks, geese, goats, pigs, reindeer, sheep, turkeys, and water buffalo. Fish account for 6 percent of the total world supply of protein and about 24 percent of animal protein if the use of fishmeal in animal feeds is counted (FAO 1981).

Although humanity uses domesticated animals for food, transportation, work, and various industrial products, only about fifty species have been domesticated. The most important domesticated animals came from the same cradles of civilization as the major staple

Table 3. Plants and Animals as Part of the Human Diet, 1983–1985 (Per Capita Per Day)

Type	Calories		Protein (grams)		Fat (grams)	
	Plants	Animals	Plants	Animals	Plants	Animals
Developed Nations	2,364	1,010	42.4	54.9	48.5	78.6
Developing Nations	2,222	202	46.9	11.3	27.6	15.2
World	2,258	407	45.8	22.4	32.9	31.3

Source: FAO 1987a.

foods. Like crops, these animals have diversified to form a variety of genetically distinct types. For example, some 140 distinct breeds of European cattle *(Bos taurus)* can be found throughout the world, and most breeds comprise genetically distinct populations. Eight breeds of European cattle (the Criollos) occur in tropical Latin America, and these form some 31 distinct populations (de Alba 1987). Other cattle, such as the zebu cattle *(Bos indicus)* of India and the banteng *(Bos javanicus)* of southeast Asia, are considered different species, though they can be hybridized with European cattle.

Of the forty nations with the highest percentage of animal protein supplies derived from fish, thirty-nine are developing countries.

Wild animals contribute only a small proportion of the human diet on a global basis, but both regionally and locally, wild species' importance is often much greater. In at least seven southeast Asian countries, more than half of the animal protein consumed in the mid-1970s was derived from fish (Darus 1983), and in one portion of the Peruvian Amazon, fish account for approximately 60 percent of the animal protein that local people eat (Dourojeanni 1985). In general, wild animal species play a greater role in the human diet in developing countries than in the developed world. Of the forty nations with the highest percentage of animal protein supplies derived from fish, thirty-nine are developing countries (FAO 1984). (Japan is thirteenth.) Also, within developing countries, the poor spend proportionately more of their household income on fish than on other meat products (FAO 1981, James 1984).

In many parts of the world, wild terrestrial animals may contribute substantially to the local economy. In South Africa, Zimbabwe, and Zambia, game ranching—the controlled harvest of wild or semi-domesticated animals often confined on fenced ranches—is becoming increasingly popular. Some 8,200 landowners in South Africa earn between $28.6 million and $33.6 million annually through game ranching (Benson 1986).

Although the major food crops in use today were domesticated more than 2000 years ago, the potential for other species to play increasingly prominent roles in world agriculture is great (Vietmeyer 1986). Numerous locally important species of wild and domesticated plants could be cultivated and used much more widely (NRC 1975, Arora 1985, Haq 1988). For example, quinua *(Chenopodium quinoa)*, a staple grain of the ancient Incas, is little known outside of the highlands of Bolivia, Chile, Ecuador, and Peru, yet it is one of the world's most productive sources of plant protein (NRC 1975). Its more widespread use in tropical countries could significantly increase grain productivity and help alleviate the pressures forcing continuing agricultural expansion onto marginal lands. Similarly, several leguminous crops that tropical and subtropical peoples have cultivated for millennia are now being investigated as "new" crops for harsh tropical environments and marginal arid lands. These include yam bean *(Pachyrhizus spp.)*, marama bean *(Tylosema esculentum)*, bambara groundnut *(Voandzeia subterranea)*, jackbean and swordbean *(Canavalia spp.)*, and winged bean *(Psophocarpus tetragonolobus)* (NRC 1979). Many wild species of known value are also likely to become more important as they become domesticated. Palm hearts, for example, are a valuable product still harvested largely from wild species of the family Palmae in the neotropics; domesticating these species or developing sustainable extractive harvesting systems could create valuable industries in many tropical communities (NRC 1975).

The size of the untapped store of locally important food species that may play significant roles on wider scales in the future is suggested by the number of wild species that only indigenous peoples eat. For example, one tribe in Brazil uses 38 different wild species of trees for food (Prance et al. 1987). Often, knowledge of how to use specific species eludes scientists or even some groups within the local society. One survey in Sierra Leone found that local women could name thirty-one products gathered or made from wild species while the men could only name eight (FAO/SIDA 1987).

As with plants, the use of edible animal species can be expanded in two ways—by using species of local importance more widely and by domesticating wild species. In Asia, for example, many locally important domesticated species of cattle and pigs—including the banteng, mithan *(Bos frontalis)*, yak *(Bos grunniens)*, and Sulawesi warty pig *(Sus celebensis)*—may be well-suited to environmental conditions in other tropical and sub-tropical countries (NRC 1983). Asia also contains several wild species of pigs and cattle that could be valuable domesticates or—as in the case of the kouprey *(Bos sauveli)* and the wild banteng *(Bos javanicus)*—breeding stock for already-domesticated species (NRC 1983). Several of these wild cattle, including the kouprey, currently face extinction.

Many other animals hold promise as semi-domesticated or domesticated species. In Central America, a semi-domesticated population of the green iguana *(Iguana iguana)*, endangered in much of its range, has been established in Panama, and the Iguana Management Project will try to establish a similar population in Costa Rica (Chapin 1986, Ocana et al. 1988). Sustainably harvesting iguana could provide valuable protein to local communities and help ensure the survival of the tropical forest habitat that the iguana requires.

Medicines

Many of the world's medicines contain active ingredients extracted from plants, animals, or microorganisms or synthesized using natural chemicals as models. Tropical species have been particularly important sources of medicines, in part because they contain a wide array of toxic compounds that have evolved to hinder herbivory or predation and many active medical compounds are derived from such toxins. Just three of the many important medicinal tropical plant species are serpent-wood *Rauvolfia serpentina* (anti-hypertensive drugs), Mexican yams *Dioscorea composita* (steroids), and the rosy periwinkle *Catharanthus roseus* (anti-cancer drugs).

The medical benefits of the world's biodiversity are not limited to plant compounds. A wealth of antimicrobial, antiviral, cardioactive, and neurophysiologic substances have been derived from poisonous marine fauna (Ruggieri 1976), and the venoms of various arthropods have medicinal potential (Bettini 1978). Domesticated animals have provided humanity with hormones and enzymes, and animals have also played important roles in behavioral studies and medical research. A long list of primates, including baboons *(Papio spp.)*, chimpanzee *(Pan troglodytes)*, and the African vervet monkey *(Cercopithecus aethiops)* are invaluable for biomedical research on AIDS and other diseases. Fungi and microbes have provided humanity with such life-saving drugs as the antibiotics, penicillin and tetracycline, and the immunosuppressant cyclosporin, which has greatly increased the survival rate for heart- and kidney-transplant patients (Byrne 1988).

New medicinal compounds are often derived from species that have been used as folk remedies for centuries. In one study of 119 plant-derived drugs used in Western medicine, some 77 percent were found to have been used in folkloric medicine by indigenous cultures (Farnsworth 1988), and one of the most promising new anti-malarial drugs—*qinghaosu*—is the active ingredient of a Chinese herbal medicine used for centuries to treat malaria and rediscovered only in 1971 (Wyler 1983). Of course, some of the most threatening modern diseases, including AIDS and some types of cancer, have historically had little impact on traditional societies; in such cases, traditional medicinal species may have no more than other species to offer.

While it is reasonable to turn to species-rich regions to find promising medicinal compounds, will that search be made? In the 1960s and 1970s, pharmaceutical companies became increasingly reliant on the chemical synthesis of drugs, and the search for natural chemicals slowed. At the same time, efforts to locate natural chemicals shifted from plants and animals to microorganisms and fungi because they can be collected, cultured, and screened cheaply and easily and can be readily grown in laboratories to produce commercial quantities of active compounds.

If they can't patent the use of a natural compound, companies have no incentive to make the long-term investments required for screening and developing them.

Private industry's declining interest in sampling plants and animals for natural medicinal compounds may have been further weakened by uncertainty over property rights for natural medicines (Farnsworth 1988, Sedjo 1989). If they can't patent the use of a natural compound, companies have no incentive to make the long-term investments required for screening and developing them. Precedents do exist for such patents: Eli Lilly Co., for example, has exclusive rights to the anti-cancer agent vincristine derived from the rosy periwinkle (Farnsworth 1988). But synthetic products have advantages: clear property rights and secure supplies.

Despite these drawbacks, wild species remain vitally important to medicine's future. About four fifths of the people in developing countries still rely on traditional medicines for health care (Farnsworth 1988). Even in the high-tech medical industries of developed countries, wild species are unlikely to lose their prominence, and natural compounds still account for a significant portion of medicines in developed countries. Between 1959 and 1973, one fourth of all prescriptions dispensed in the United States contained active ingredients extracted from vascular plants (Principe, in press). Most of the synthetic drugs that have become increasingly important are modeled after natural products, so their use should not slow the search for natural compounds. Indeed, very recently, interest in that search has revived (Sedjo 1989), thanks in part to new arrangements whereby pharmaceutical firms, collectors, and the nation of species origin can share profits from the discovery and use of medicinal plants, animals, and microbes (Sedjo 1989).

Finding new natural drugs is clearly in society's interest, and declining private investment has been partially offset in some countries by increased public expenditure. The U.S. National Cancer Institute, for example, launched an $8-million program in 1987 to screen 10,000 natural substances each year for five years against one hundred cancer cell lines and the AIDS virus (Booth 1987). The People's Republic of China, Japan, India, the Federal Republic of Germany, and other countries also have active research programs directed at natural drugs (Farnsworth 1988).

The search for new medicines presents a tremendous opportunity for tropical countries. Relatively low-cost facilities could be established in developing countries to initially screen compounds (Eisner, in press). Such facilities would be labor-intensive and relying on them would have several advantages over shipping materials to other countries: the chance of locating active compounds would be increased since fresh materials could be tested; local industries would be developed; in-country scientific expertise would be increased; and medicines could be developed to treat the most troublesome regional diseases, such as malaria and schistosomiasis. Such local institutions could help fill a gap left by biomedical research in the developed world, which is tailored toward the needs of people in developed countries.

Industrial Uses

Plants and animals are major industrial feedstocks. Plants provide such products as natural rubber, waxes, oils, fuelwood, timber, forage, fibers, resins, and they are also used for ornamental purposes. Animals provide oils, fuel, silk, and feathers, and throughout the world they are valued as pets. Everywhere, local species tend to be the primary source of industrial products important to the region, regardless of its relative species-richness. But, taking a global perspective, species-rich ecosystems will no doubt provide the most products for the future use of collective humanity.

The most significant industrial uses of the world's biota trace back to the world's forests. Global timber trade amounts to some $77 billion annually (FAO 1989). In addition, almost half of the world's population, more than 2 billion people, use wood as their primary source of energy (WRI/IIED 1988), while in sub-Saharan Africa wood accounts for 80 percent of the total energy consumed (WEC 1987). Non-timber forest products are also extremely important in many countries, and their value may sometimes exceed that of traditional timber products. In 1986, Indonesia earned $86 million from exports of rattan (Vatikiotis 1989), and in various parts of the world bamboo, brazil nuts, rubber, and fruit are mainstays of local economies and significant exports. The value of non-wood forest products is much greater than generally assumed. One recent study of a tropical forest in Peru found the value of non-wood forest products to be two to three times that of the timber (Peter et al. 1989).

Particularly in tropical countries, many of the benefits derived from forests are threatened by the loss or degradation of forest ecosystems. Between 1976 and 1980, an estimated 7.3

million hectares of closed tropical forest was converted to other land uses each year and an additional 4.4 million hectares annually was logged (Lanly 1982, Melillo et al. 1985). Conserving these forest ecosystems is a high priority from the standpoint of maintaining biological diversity as well as from that of meeting the needs of people who depend upon the forest resources.

With rapid population growth in the tropics and the loss and degradation of tropical forest resources, tree species are needed for intensive forest cultivation to meet the growing demand for tropical timber products, for fuelwood production and agroforestry schemes in tropical countries, and for stabilizing and restoring degraded tropical soils. Since the species that will fill these needs in tropical countries will be primarily tropical in origin, the protection of the diversity of tropical species is of high priority.

The combined potential of plants, animals, and microorganisms in the war against hunger, disease, and economic stagnation is only beginning to be tapped.

Various locally important tropical species are now being investigated for possible use over larger regions of the tropics (Vietmeyer 1986). In particular, leguminous species have received special attention because they can fix nitrogen and this property is essential to agroforestry and land restoration. Leucaena (*Leucaena leucocephala*), which originated in Mexico and was widely utilized and dispersed by the Mayan and Zapotec civilizations, is now being introduced into numerous tropical nations. Leucaena can produce nutritious forage, firewood, timber, and organic fertilizer, and it can be used to revegetate degraded lands, establish windbreaks, and provide shade (NRC 1977). Similarly, such species as mesquite *(Prosopis spp.)* and a variety of species of acacias are now being examined for use in rehabilitating degraded lands, as a component of agroforestry, or as plantation species (NRC 1979). The combined potential of plants, animals, and microorganisms in the war against hunger, disease, and economic stagnation is only beginning to be tapped.

IV. Extinction: How Serious is the Threat?

The world's biological diversity has co-evolved with human culture. Humanity has applied growing knowledge and skills to order and manipulate nature to meet changing human needs. In this process, people have hunted, fished, and gathered species for food, fuel, fiber and shelter. They have eliminated competing or threatening species, domesticated plants and animals, cut forests, used fire to alter habitats, and recently even significantly changed hydrological and geochemical cycles. As a result, the landscape and, to a lesser extent, the sea, today reflect human culture.

These various human impacts on the biota may increase or decrease the genetic, species, and habitat diversity in a particular region. But the most profound—and irreversible—impact of human activities on the biosphere is the substantial acceleration of species extinction. Today, human beings use, divert, or waste about 40 percent of the total terrestrial net primary productivity, or about 25 percent of the world total (Vitousek et al. 1986). Not surprisingly, humanity's impact on the remainder of the world's biota is also substantial.

Trends in Species Extinctions

Humanity's impact on species extinction rates goes back several thousand years, but over the last century the human factor has increased dramatically. *(See Box 3.)* Written records of extinctions are most complete for birds and mammals. However, these data must be interpreted with considerable caution. Although some species listed as extinct may subsequently be rediscovered, recorded extinctions tend to underestimate the true number of extinctions because the status of many species, particularly in tropical forests, is unknown (Diamond 1988b). This problem notwithstanding, the rate of known extinctions of birds and mammals increased fourfold between 1600 and 1950. *(See Figure 6.)* By 1950, recorded extinction rates had climbed to between one-half and one percent of birds and mammals per century. Since 1600, some 113 species of birds and 83 species of mammals are known to have been lost. *(See Table 4.)* Between 1850 and 1950, the extinction rate of birds and mammals averaged one species per year. The rapid growth in the rate of species extinction is a telling measure of the status of the world's biodiversity.

Current rates of extinction among birds and mammals are perhaps 100 to 1000 times what they would be in unperturbed nature.

How does the current rate of extinction compare with average extinction rates in the absence of human influences? For most groups

Box 3. A History of Extinction

Extinction rates have varied considerably over the history of life on earth. Paleontologists distinguish five episodes of ''mass extinctions''—relatively short (1 million to 10 million year) periods during which a significant fraction of diversity in a wide range of taxa went extinct. The most significant mass extinction, at the end of the Permian (250 million years ago), may have eliminated 77 to 96 percent of species (Valentine et al. 1978, Raup 1979). Even apart from these mass extinctions, background rates of extinction are not constant. For example, for the past 250 million years, relatively high rates of extinction have occurred nine times—at intervals of approximately 26 to 28 million years. Two of these nine episodes were mass extinctions, one in the late Triassic, 220 million years ago, and one in the late Cretaceous, 65 million years ago (Sepkoski and Raup 1986).

Global biological diversity is now close to its all time high.

Global biological diversity is now close to its all time high (Wilson 1988b). Floral diversity, for example, reached its highest level ever several tens of thousands of years ago (Knoll 1986). Similarly, the diversity of marine fauna has risen to a peak in the last few million years (Raup and Sepkoski 1982).

Humanity's first significant contribution to the rate of global extinction may have occurred 15,000 to 35,000 years ago, when hunting of large mammals apparently caused or contributed to significant extinctions in North and South America and Australia (Martin 1973, 1986). These three continents

lost 74 to 86 percent of the genera of ''megafauna''—mammals greater than 44 kg—at that time. While the cause of these extinctions remains a matter of controversy *(see Martin and Klein 1984)*, even if humanity is not wholly responsible, there is no doubt that for millennia, people have significantly altered the landscape with untold effects on native flora and fauna. For at least 50,000 years, intentional burning has occurred in the savannas of Africa (Murphy and Lugo 1986). At least 5,000 years ago, in Europe, deforestation and the conversion of wildlands to pasture began and there is evidence in North America that for as long as 4,000 years indigenous peoples influenced the structure of forest communities, provided opportunities for weedy species and such herbivores as bison *(Bison bison)* to expand their ranges, and caused at least local species extinctions (Delcourt 1987). In Central America, forest had already been removed from large areas before the Spanish arrived (D'Arcy 1977).

The prehistoric colonization of islands by human beings and their commensals substantially affected the diversity of island species. Fossil evidence suggests that 98 species of endemic birds were present in the Hawaiian Islands in A.D. 400 when the islands were first colonized by Polynesians. About fifty of these species became extinct before the first European contact in 1778. Most experts believe that these extinctions resulted from a combination of the clearing of extensive tracts of lowland forest for agriculture, predation and disturbance by introduced species (the Polynesian rat, domestic pig, and domestic dog), and hunting (Olson and James 1982, 1984; Vitousek 1988; Olson 1989). Similarly, following human colonization of New Zealand in A.D. 1000, the introduction of the domestic dog and the Polynesian rate, combined with the

Box 3. (cont.)

deforestation of large areas by fire and intensive hunting of larger birds, led to the extinction of 13 species of moas (large flightless birds) and 16 other endemic birds before the arrival of Europeans (Cassels 1984). Humanity is thought to have caused other extinctions following the colonization of Madagascar in A.D. 500 and the Chatham Islands in A.D. 1000. Early human colonization of oceanic islands may have led to the extinction of as many as one-quarter of the bird species that existed several millennia ago (Olson 1989).

In the 15th and 16th centuries, the global spread of European cultures and their attendant livestock, crops, weeds, and diseases increased the loss of island flora and fauna and added to the threats to continental species (Crosby 1986). In later centuries, the growth of trans-oceanic human travel and commerce led to the spread of a tremendous variety of species to new regions of the world and to the human colonization of many uninhabited islands. Between 1840 and 1880, more than 60 species of vertebrates were released in Australia (K. Myers 1986). Between 1800 and 1980, the number of introduced insect species in the United States grew from about 36 to more than 1200 (Simberloff 1986b).

Many of the European introductions and colonizations, like those of earlier colonizers, significantly influenced native flora and fauna. In Hawaii, the arrival of European explorers added cats, two new species of rats, the barn owl, the small Indian mongoose, and several avian diseases. In the next two centuries, habitat degradation, disease, and predation caused the loss of 17 endemic bird species (Olson 1989), reducing the endemic avifauna to 31 percent of the diversity found in A.D. 400; several more species now verge on extinction. Europeans first visited the uninhabited Mascarene islands (Mauritius, Reunion, and Rodrigues) in the early 1500s and released pigs and monkeys on the islands. In the mid-1600s, the Dutch settled the islands, and in the next 300 years twenty species of birds—including the Dodo—and 8 species of reptiles were lost (Nilsson 1983). The extreme vulnerability of island endemics is exemplified by the fate of the flightless Stephen Island wren (*Xenicus lyalli*), driven to extinction by a single cat owned by a lighthouse keeper on an islet off New Zealand (Diamond 1984).

of organisms, the average "lifespan" of a species is on the order of 1 million to 10 million years (Raup 1978), so only 1 to 10 species out of the *total current biota* of some 10 million species would be expected to be lost each year. These estimates are most accurate for widespread species and particularly for marine species because their fossil record is most complete. But estimates for terrestrial mammals also fall within this range (1 million to 2 million years; Raup and Stanley 1978). Among organisms with restricted ranges, prehistoric extinction rates may have been higher than the averages suggest, though it is safe to say that for widespread species, the so-called background extinction rate is *extremely* small, indeed close to zero. Among the approximately 13,000 species of birds and mammals on earth today, an extinction would be expected only every 100 to 1000 years. Current rates of extinction among birds and mammals are thus perhaps 100 to 1000 times what they would be in unperturbed nature.

Besides birds and mammals, marine invertebrates are the only other group for which the data in Table 4 present an even approximately accurate representation of the true number of

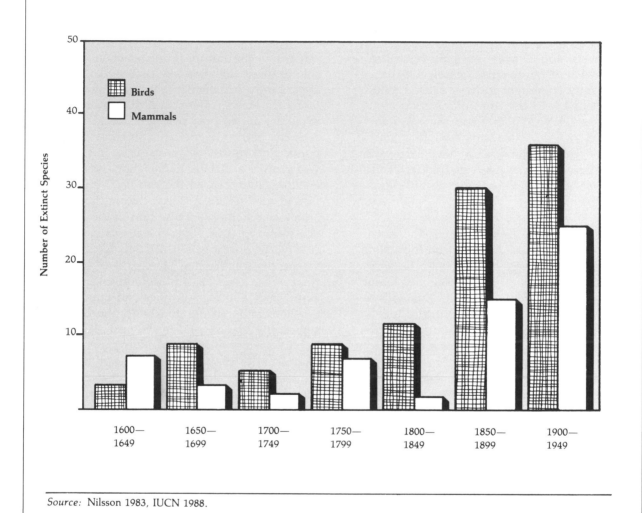

Figure 6: Number of Species of Birds and Mammals that Became Extinct Between 1600 and 1950

Number of Extinct Species

☒ Birds
☐ Mammals

1600—1649 1650—1699 1700—1749 1750—1799 1800—1849 1850—1899 1900—1949

Source: Nilsson 1983, IUCN 1988.

As far as can be determined, there has been but a single extinction of a marine invertebrate during recorded history.

extinctions. As far as can be determined, there has been but a single extinction of a marine

invertebrate during recorded history (a limpet associated with eel-grass in north-eastern United States) (G. Vermeij, University of California, personal communication, March 1989). Among other groups of species, the number of recorded extinctions probably represent only a small fraction of the total since most species have not even been described and the status of described species is generally poorly known. Recent surveys for freshwater fish thought to still exist on the

Table 4. Recorded Extinctions, 1600 to Present

Taxa	Mainland[a]	Island[b]	Ocean	Total	Approximate Number of Species	Percentage of Taxa Extinct Since 1600
Mammals	30	51	2	83	4,000	2.1
Birds	21	92	0	113	9,000	1.3
Reptiles	1	20	0	21	6,300	0.3
Amphibians	2	0	0	2	4,200	0.0
Fish[c]	22	1	0	23	19,100	0.1
Invertebrates[c]	49	48	1	98	1,000,000+	0.0
Vascular Plants[d]	245	139	0	384	250,000	0.2
Total	370	351	3	724		

a. Landmasses greater than 1 million square kilometers (the size of Greenland and larger).

b. Landmasses less than 1 million square kilometers.

c. Totals primarily representative of North America and Hawaii *(See Ono et al. 1983)*.

d. Vascular taxa (includes species, subspecies, and varieties).

Sources: IUCN 1988 (fish and invertebrates); Nilsson 1983, supplemented by species listed as extinct in IUCN 1988 (vertebrates other than fish); Threatened Plants Unit, World Conservation Monitoring Centre, 15 March 1989, personal communication (plants); G. Vermeij, personal communication (marine invertebrates); Wilson 1988a (number of species).

Malay peninsula failed to locate 55 percent of the 266 species known from the region (Mohsin and Ambak 1983).

A useful rule of thumb is that if a habitat is reduced by 90 percent in area, roughly one-half of its species will be lost.

Predicting future extinction rates is clearly even more difficult than estimating the current rate. Such predictions are generally based on projected rates of habitat loss and the relationship between species richness and habitat area (known as a "species-area curve"). A useful rule of thumb is that if a habitat is reduced by 90 percent in area, roughly one-half of its species will be lost. *(See Figure 7.)* Using the species-area approach, Simberloff (1986a) found that deforestation in neotropical moist forests between 1986 and the turn of the century could extinguish about 15 percent of plant species in the neotropics and 12 percent of bird species in the Amazon basin. In theory, if deforestation were to continue until all forest except that legally protected from harvest is eliminated, 66 percent of plant species and 69 percent of bird species would be lost.

Simberloff's approach can be extended to cover the effects of deforestation on species extinction rates in the remainder of the tropics. *(See Appendix 2—Calculating Extinctions due to Deforestation.)* If current trends continue, by

Figure 7: Species-Area Curve

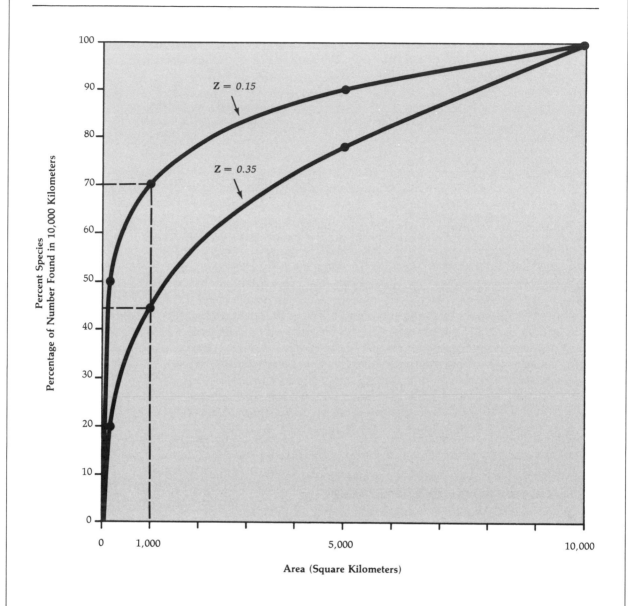

The number of species found in a region increases with the area of habitat in a predictable manner. Consequently, it is possible to predict the effect of habitat loss on the number of species found in a region with use of a species-area curve such as this. A useful rule of thumb is that if a habitat is reduced by 90 percent in area, roughly one-half of its species will be lost. The two curves show the difference in the relationship found in island (z = 0.35) and continental (z = 0.15) habitats. (See Appendix 2.)

2020 species loss could amount to 6 to 14 percent of species in closed tropical forests in Africa, 7 to 17 percent of species in Asia, and 4 to 9 percent of species in Latin America. *(See Table 5.)* If actual deforestation rates are double the estimated rates, the potential species loss would increase by 2 to 2.5 times. Globally, this would amount to the extinction of roughly 5 percent of plants and 2 percent of birds at current rates of deforestation, or approximately 10 percent of plants and 5 percent of birds at twice the estimated rate of deforestation. Between 50 and 90 percent of arthropods occur in tropical forests; if they are distributed among tropical regions in Asia, Africa, and Latin America in the same proportions as vascular plants, then approximately 5 percent of the world's arthropods would be lost if current rates of deforestation continue and 15 percent if those rates double.

Because arthropods and plants account for much of the world's species diversity, the potential extinction of species in these groups roughly indicates the effect of tropical forestation on the world's entire species diversity. Between 1990 and 2020, species extinctions caused primarily by tropical deforestation may eliminate somewhere between 5 and 15 percent of the world's species. With roughly 10 million species on earth, this would amount to a potential loss of 15,000 to 50,000 species per year or 50 to 150 species *per day.*

Between 1990 and 2020, species extinctions caused primarily by tropical deforestation may eliminate between 5 and 15 percent of the world's species. With roughly 10 million species on earth, this would amount to a potential loss of 15,000 to 50,000 species per year or 50 to 150 species **per day.**

This analysis generally agrees with other predictions of species extinction rates that have been made using a variety of techniques. *(See Table 6.)* These studies indicate that if current trends continue, roughly 5 to 10 percent of the world's species will be lost per decade over the

Table 5. Predicted Percent Loss of Tropical Forest Species due to Extinction

Region	Historical to 1990	1990–2000	1990–2010	1990–2020	1990–2020 with 2x Deforestation	Only Legally Protected Areas Remaining
Africa and Madagascar	10–22	1–2	2–4	3–7	6–14	35–63
Asia and Pacific	12–26	2–5	4–10	7–17	22–44	29–56
Latin America and Caribbean	4–10	1–3	3–6	4–9	9–21	42–72

Notes: Estimates based on species-area model ($0.15 < z < 0.35$). Forest loss based on projections by Lanly (1982) for the period 1981 to 1985. *See Appendix 2 for discussion.*

Table 6. Extinction Rate Estimates

Estimate	Global Loss Per Decade (Percent)	Method of Estimation	Source
15–20 percent of species between 1980 and 2000	8–11	Estimated species-area curve; forest loss based on Global 2000 projections	Lovejoy (1980)
17,500 species per year lost in tropical rain forests	2[a]	Half of rain forest species assumed to be local endemics becoming extinct with forest loss	Wilson (1988)
Committed to loss of **12 percent of plant species** in neotropics and **15 percent of bird species** in Amazon basin by 2000	—[b]	Species-area curve ($z = 0.25$)	Simberloff (1986)
25 percent of species between mid-1980s and 2015	9[c]	Number of species occurring in tropical forested areas likely to be deforested or severely disturbed by 2015; half of species in these areas assumed to be lost	Raven (1988a, b)
5–15 percent loss between 1990 and 2020	2–5[d]	Species-area curve ($0.15 < z < 0.35$); forest loss 1–2 times FAO projection for 1980–1985.	This study

a. Estimate applies only to immediate extinction of local endemics and not to the slower loss of widespread species resulting from insularization of habitat.

b. Extinction estimates apply to the equilibrium number of species if forest loss continues to year 2000 and then halts. How long it will take for this equilibrium to be achieved is not known.

c. Extinctions will take place during this time period (mid-1980s to 2015) or shortly thereafter.

d. Extinctions will take place during this time period (1990 to 2020) or shortly thereafter. The slow loss of species in forest fragments is not included in this model.

next quarter century. This rate of extinction would be unparalleled since the last mass extinction event at the end of the Cretaceous Era, 65 million years ago.

If current trends continue, roughly 5 to 10 percent of the world's species will be lost per decade over the next quarter century. This rate of extinction would be unparalleled since the last mass extinction event at the end of the Cretaceous Era, 65 million years ago.

During the past century, the pattern of species extinction has changed in two ways. Historically, extinction threatened mainly island-dwelling species and a handful of vulnerable continental species: some 74 percent of extinctions of birds and mammals were of island-dwelling species. *(See Table 4.)* Currently, though island species remain highly endangered, 66 percent of endangered and vulnerable terrestrial vertebrates are continental. *(See Table 7.)* In addition, habitat loss has become the main threat to species survival. Historically, species introductions and over-exploitation were equally important influences. *(See Tables 8 and 9.)*

The Geography of Extinction

Species in some habitats and biomes are more threatened with extinction than species in others are. These seriously threatened geographic regions are often referred to as ''critical ecosystems.'' Most extinctions in the coming decades will occur on islands and in closed tropical forests. Regions with a Mediterranean climate also contain large numbers of threatened plants and invertebrates, and fish and aquatic invertebrates are seriously threatened in

Table 7. Endangered and Vulnerable Species

Taxa	Mainland[a]	Island[b]	Ocean	Total
Mammals	159	48	9	216
Birds	91	87	0	178
Reptiles	41	21	6	68
Amphibians	14	0	–	14
Fish[c]	193 (443)[d]	21	0	214 (464)[b]
Invertebrates (Class Insecta)[c]	138	239	0	377
Invertebrates (Others)[c]	233	99	2	334
Vascular Plants[e]	3985	2706	0	6691

a. Landmasses greater than 1 million square kilometers (size of Greenland and larger).

b. Landmasses less than 1 million square kilometers.

c. Primarily representative of North America and Hawaii.

d. Including 250 species of cichlids endangered in Lake Victoria.

e. Vascular taxa (includes species, subspecies, and varieties).

Source: IUCN 1988; Threatened Plants Unit, World Conservation Monitoring Centre, 15 March 1989, personal communication.

Table 8. Causes of Extinction

Group	Habitat Loss	Over-Exploitation[a]	Species Introduction	Predator Control	Other	Unknown
		Percent Due to Each Cause				
Mammals[b]	19	23	20	1	1	36
Birds[b]	20	11	22	0	2	37
Reptiles[b]	5	32	42	0	0	21
Fish[c]	35	4	30	0	4	48

a. Includes commercial, subsistence and sport hunting and live animal capture for pet, zoo, and research trades.

b. Value reported represents the percentage of species whose extinction was caused *primarily* by the factor indicated.

c. Value reported represents the percentage of species whose extinction was *influenced* by the factor indicated; thus, row exceeds 100 percent.

Sources: Day 1981; Nilsson 1983; Ono et al. 1983.

lakes and rivers throughout the world (Mooney 1988, Ono et al. 1983). This geographical pattern of species endangerment is reflected in the distribution of endangered species in the United States. Half of the U.S. plant species in danger of extinction within the next decade occur in only three states or territories—Hawaii and Puerto Rico (islands with tropical forests) and California (a Mediterranean-climate zone) (CPC 1988). Almost 80 percent of endangered and threatened fish in the United States are confined to the arid regions of the southwest (primarily desert fishes) and the southeast (primarily warm water river species) (Ono et al. 1983).

Freshwater Ecosystems

Freshwater ecosystems are the terrestrial analog of oceanic islands. Many freshwater invertebrates and fishes are found only in individual lakes, rivers, or portions of rivers because waterfalls and other barriers have limited their ranges. Freshwater species are thus particularly threatened by habitat loss. Moreover, freshwater communities are extremely susceptible to

extinctions caused by the introduction of exotic species, since many of these communities have developed in the absence of various predators and parasites. Collectively, these threats can devastate freshwater faunas. On the island of Singapore, 34 percent of 53 species of freshwater fish collected in 1934 could not be located in exhaustive searches only 30 years later (Mohsin and Ambak 1983).

Species introductions threaten aquatic communities throughout the world. The introduction of lamprey helped extinguish several species of cisco (*Coregonus spp.*) and has endangered three other ciscos in the Great Lakes of North America (Ono et al. 1983). Largemouth bass (*Micropterus salmoides*) introduced in Lake Atitlán have seriously reduced populations of two small native fish species *Poecilia sphenops* and *Cichlasoma nigrofasciatum,* and the peacock bass (*Cichla ocellaris*) introduced into Gatun Lake in Panama eliminated six of the eight previously common fish species and drastically reduced a seventh (Zaret and Paine 1973). In California, where over one third of the freshwater fish fauna is made up of non-native

40

Table 9. Threats to Species

Group	Percent Due to Each Cause[a]				
	Habitat Loss	Over-Exploitation[a]	Species Introduction	Predator Control	Other
Mammals	68	54	6	8	12
Birds	58	30	28	1	1
Reptiles	53	63	17	3	6
Amphibians	77	29	14	–	3
Fish	78	12	28	–	2

a. Value represents the percent of species whose endangerment is influenced by each factor; row exceeds 100 percent. Threatened species are those species and subspecies considered by IUCN to be globally endangered, vulnerable or rare.

b. Includes commercial, subsistence and sport hunting and live animal capture for pet, zoo, and research trades.

Source: Prescott-Allen and Prescott-Allen 1978.

species, nearly 20 percent of the 66 native species are endangered (Mooney et al. 1986), and the introduction of the nile perch (*Lates niloticus*) into Lake Victoria has substantially reduced the fish harvest from the lake and helped endanger the more than 250 species of cichlid fish in the lake (Barel et al. 1985, IUCN 1988). Species introductions have contributed to 30 percent of the extinctions of fish.

Habitat loss has been an even greater threat than species introductions to freshwater communities, especially in arid regions where the human use of water may drastically alter substantial portions of the natural aquatic habitat. The construction of dams, drainage of wetlands, channelization of streams, and capping or tapping of springs has endangered many freshwater communities. In the southeastern United States, 40 to 50 percent of the freshwater mollusk (snail) species are now extinct or endangered due to the impoundment and channelization of rivers (IUCN 1983). Species are threatened not only by changes in water regimes but also by associated changes in water temperature. The release of deep cold

water from behind dams into the Colorado River, for example, has threatened the Colorado squawfish (*Ptychocheilus lucius*) (Williams et al. 1985). Human needs and the requirements of aquatic organisms conflict most in desert regions, where many aquatic species may be endemic to small springs threatened by groundwater pumping or surface water use. In desert areas of North America, 46 fish taxa are classified as endangered, 18 taxa are extinct, and several taxa have been saved from extinction in coffee cans or water troughs when their habitats were pumped dry (Williams et al. 1985).

Islands

Island species remain among the most critically threatened components of the world's biological diversity. *(See Table 10.)* Some 10 percent of Hawaii's native flora is now extinct, and fully 40 percent is threatened. Similarly, 8 percent of the endemic flora of Mauritius is extinct and 30 percent is endangered (Davis et al. 1986). In general, species occurring on islands historically free of most predators or

Table 10. Status of Island Plant Species

Island	Number of Native Species	Number of Endemic Species	Percent Endemic	Percent Threatened Endemics[a]	Threatened Endemics
Ascension	25	11	44	9	82
Azores	600	55	9	23	42
Canary Islands	–	500	–	377	75
Galápagos	543	229	42	135	59
Hawaii[b]	970	883	91	c.353	c.40
Juan Fernandez	147	118	80	93	79
Lord Howe Island	379	219	58	70	32
Madeira	760	131	17	86	66
Mauritius	c.850	c.280	c.33	119	c.42
New Caledonia	3250	2474	76	146	6
New Zealand	2000	c.1620	c.81	132	c.8
Norfolk	174	48	28	45	94
Rodrigues	145	40	28	36	90
Saint Helena	60	50	83	132	8
Socotra	–	215	–	130	60

a. Includes IUCN categories "endangered," "vulnerable," and "rare."

b. Species number and endemism from Gentry 1986.

Source: Davis et al. 1986.

pests have been seriously affected by the ecosystem changes resulting from the arrival of human beings and the rats, cats, and other species that follow them around.

A particularly dramatic example of threatened island biodiversity is Madagascar. This island boasts one of the world's most distinctive assemblages of plants and animals. Some 93 percent of Madagascar's 28 primate species, 80 percent of plant species, 54 percent of the 197 native breeding birds, and 99 percent of the 144 amphibians are endemic to the island (Davis et al. 1986, Jenkins 1987). Already, the extinction of species on Madagascar represents a substantial loss both to the Malagasy people and to the global community. Since human beings first arrived on the island around A.D. 500, 14 species of lemurs, a pygmy hippopotamus, two species of land tortoise, and seven species of

elephant birds have been lost (including the largest bird species ever known to have existed, measuring some 3-meters tall and weighing an estimated 450 kilograms) (Nilsson 1983).

Nearly 80 percent of Madagascar's natural vegetation has been altered, largely through slash-and-burn agriculture, and in some portions of the country very little natural vegetation remains. Only 1.5 percent of the island is currently protected, so this habitat disturbance could contribute substantially to species extinction rates. According to the species-area approach, an 80-percent decline in the extent of natural habitat could cause the loss of up to 45 percent of the original species. Currently, 6 of the 28 primate species (and 2 subspecies) and 5 of the 106 endemic bird species are listed as endangered or vulnerable by the IUCN (IUCN 1988, Jenkins 1987).

Spurred by the threat of losses of this magnitude, the Government of Madagascar has prepared an environmental action plan that would increase the number of protected areas from 36 to 50, upgrade protection and management capacity, and establish integrated conservation/development projects in communities adjacent to the protected areas. Over twenty years, implementing this plan will require approximately $120 million. This sum, only about one third of the 1987 development assistance provided to Madagascar, is a tiny cost for an activity central to the long-term integrity of Madagascar's environment.

Mediterranean Climate Zones

Five geographic regions have a Mediterranean climate characterized by cool wet winters and warm dry summers—parts of California and Chile, the Mediterranean basin in southern Europe and North Africa, the Cape regions of South Africa (the *fynbos*), and southwestern Australia. The plant species richness and endemism in these regions rival those of the tropics. The four regions shown in Table 11 contain roughly 20 percent as many plant species as the species-rich closed tropical forests (9 percent of the global total of plant species) but cover less than 20 percent of the area (1.1 percent of the earth's land surface). The average range of plant species is as small or smaller in Mediterranean regions than in the tropics, and plant species richness per unit area in Mediterranean regions equals or exceeds that of the tropics (Gentry 1986).

Habitats in Mediterranean-climate regions have been among the most extensively disturbed in the world and have been further degraded by introduced species. Some 33 percent of the *fynbos* vegetation has been lost in South Africa, with the proportion in lowland regions reaching 70 percent loss. In California,

Table 11. Plant Species Richness and Level of Endangerment in Mediterranean-climate Regions

Region	Area (million hectares)	Number of Plant Species	Percent Endemic	Threatened Taxa[a]	Percent Threatened
Cape Region, South Africa	7.5	8,578	68	529	6.2
California	41.1	5,046	30	604[b]	12.0
S.W. Australia	32.0	3,600	68	554[c]	15.4
Chile[d]	75.2	5,500	50	110	2.0
Total	155.8	22,724	54	1797	7.9

a. Includes IUCN categories: "endangered," "vulnerable," and "rare;" these species are globally threatened.

b. S.W. Australia's classification of species as globally "rare or endangered" does not directly correspond to IUCN categories described in footnote "a".

c. Estimate for Southwestern province of Australia based on number of endangered and vulnerable species listed for province (853) minus the number insufficiently known (259) and extinct (40) (Leigh et al. 1982).

d. Data include regions without Mediterranean climate (primarily desert and temperate rain forest). Thus, both area and species numbers account for more than just the Mediterranean-climate zone.

Sources: Leigh et al. 1982; Gentry 1986; Davis et al. 1986; Mooney 1988.

nearly 10 percent of the total number of plant species are naturalized alien species. Human impacts in the Mediterranean basin itself have been so pervasive that virtually none of the original climax vegetation remains. Most remaining species are characteristic of early successional stages of the original community (Mooney 1988). As a result of extensive habitat change, the introduction of exotics, and the high degree of local endemism of plants in Mediterranean regions, an estimated 8 percent of the taxa in these regions is considered globally endangered, vulnerable, or rare. *(See Table 11.)* This figure is slightly lower than estimates provided in the next section for temperate regions more generally (up to 10 percent) because species listed as "indeterminant" and "insufficiently known" are not included in the figure for the Mediterranean zones. If they were, the threat to California would rise to 15 percent, Cape Province of South Africa to 17

percent, and Southwest Australia to 24 percent (Leigh et al. 1982, Davis et al. 1986).

Temperate Regions

In general, the temperate biota is less threatened than tropical biota by extinction, though exceptions exist in such regions as Mediterranean-climate zones and for such specific groups of organisms as freshwater species. Between 1 and 10 percent of species in temperate countries are considered by IUCN to be globally threatened, depending upon the taxa and country involved. *(See Table 12.)* Because the status of many species is poorly known, most experts consider the upper end of this range—10 percent—reasonable, barring renewed human population growth, rapid global climate change, or significant changes in land use (P. Raven, personal communication, April 1989). The Center for Plant Conservation,

Table 12. Percent of Species that are Globally Threatened in Temperate Nations[a]

Country	Mammals		Birds		Reptiles		Amphibians		Plants	
	Number Known	Percent Threatened	Number Known	Percent Threatened	Number Known	Percent Threatened	Number Known	Percent Threatened	Number Known	Percent Threatened
Canada	163	4.9	434	1.6	32	3.1			3220	0.3
United States[b]	367	10.3	1090	6.1	368	4.6	222	6.3	20000	8.5
Japan	186	4.8	632	3.0	85	2.4	58	1.7	4022	9.8
Argentina	255	10.2	927	1.9	204	3.4	124	0.8	9000	1.7
South Africa	279	7.2							23000	5.0
United Kingdom									c.1800	1.1
Belgium									c.1700	0.6
Czechoslovakia									c.2700	1.0
Finland									c.1300	0.5
Switzerland									c.2675	0.7

a. Globally threatened: includes IUCN categories "endangered," "vulnerable," "rare," "indeterminate," and "insufficiently known."

b. Includes Pacific and Caribbean Islands.

Source: WRI/IIED 1987, 1988.

for example, estimates that of the 25,000 species, subspecies, and varieties of plants native to the United States, approximately 3,000 (12 percent) are at risk of extinction in their wild habitats (CPC 1988).

Tropical Forests

Tropical deforestation will be the single greatest cause of species extinction in the next half-century. *(See Table 5.)* With concerted conservation action, it may be possible to considerably reduce this rate of extinction. But these calculations could prove to be conservative. If rates of forest loss increase significantly in the coming decades, if the deforestation of large areas fragments and degrades the natural forest habitat that remains, or if global temperatures rise rapidly, then extinction rates could greatly exceed the estimated 5 to 15 percent of tropical species by the year 2020. For example, Salati and Vose (1983) have suggested that losing a substantial portion of the Amazonian forest could drastically alter the region's hydrology and climate, potentially accelerating species loss.

Tropical deforestation will be the single greatest cause of species extinction in the next half-century.

Causes of Extinction

The extinction of a species is generally due to multiple causes. Consider, for example, what happened to the heath hen *(Tympanuchus cupido cupido).* Once widespread in the eastern United States, it had by 1890 been reduced by overhunting to 200 birds on Martha's Vineyard and in 1896 the population numbered fewer than 100. The establishment of a refuge on the island in 1908 reversed the decline in the population, and 12 years later the population had

recovered to an estimated 2,000 individuals. But, a serious fire, a hard winter, an unusually high population of predators, and an introduced poultry disease during the next five years sharply reduced the population size. By 1920, the population had become so small that genetic deterioration resulted from inbreeding, and the population eventually died in 1932 (Dawson et al. 1987).

Any factor that leads to a decline in the population size of a species makes it more vulnerable to extinction. While overhunting may have been the primary cause of the heath hen's demise, hunting alone was not sufficient. If more habitat had been available, if the exotic disease had not been introduced, or if the natural environmental fluctuations had been less severe, the heath hen could still be alive today.

The likelihood of extinction depends on which species and which threats are involved. As noted, species with small populations or restricted ranges are among those at greatest risk of extinction, though other characteristics may also increase species' susceptibility. *(See Box 4.)* Direct threats to species survival are posed by habitat loss, over-exploitation, the introduction of exotic species, and pollution, but other threats examined below—global warming and cumulative effects—can also cause extinctions.

Habitat Loss and Fragmentation

Habitat loss, degradation, and fragmentation are the most important influences upon species extinction rates. Some 67 percent of all endangered, vulnerable, and rare species of vertebrates (including fish) are threatened by habitat degradation or loss (Prescott-Allen and Prescott-Allen 1978). These factors also pose the greatest threat to invertebrates (IUCN 1983) and plants (Lucas and Synge 1978). Habitat loss has accelerated so much that in many countries relatively little natural or semi-natural habitat remains. In the Old World tropics, 49 of 61 countries have lost more than 50 percent of their original wildlife habitat, and the

Box 4. Extinction-Prone Groups of Species

Human impacts on the environment—such as habitat loss, over-exploitation, species introductions, and pollution—do not threaten all groups of species equally. At greatest risk are species with small population sizes, species whose populations vary greatly, and species with slow rates of population growth. More specifically, the following groups of organisms are particularly susceptible to extinction:

Species at higher trophic levels. Species high in the food chain tend to be large, rare animals with slow rates of population growth. Particularly susceptible to over-exploitation or habitat loss, these species include the Caribbean monk seal, great auk (*Alca impennis*), Steller's sea cow (*Hydrodamalis stelleri*), thylacine (*Thylacinus cynocephalus*, also known as the Tasmanian wolf), and many endangered birds of prey, cats, wolves, foxes, and whales.

Local endemics. Species with restricted ranges are often threatened by habitat loss. For example, the entire natural habitat of the Devil's Hole pupfish (*Cyprinodon diabolis*) is a spring measuring roughly 3 by 15 meters. Water development, pollution, or habitat alteration could easily drive the species to extinction in the wild. The extremely high rate of island species extinction further testifies to the threat that local endemics face.

Species with chronically small populations. This category overlaps the first. Since many species at higher trophic levels have sparsely distributed populations, habitat restriction or fragmentation may reduce their populations to very small levels. However, the population sizes of species at lower trophic levels may also be extremely small in a given habitat or region. Many tropical forest tree species, for instance, have very low population densities and are thus threatened by habitat loss and fragmentation.

Largest members of a guild. Whether or not a species is on the top trophic level, large species have high metabolic demands, require large habitats, and tend to occur in low densities. Thus, the largest species within a group of species sharing similar food sources (a guild) tend to be at high risk of extinction. For example, all of the lemurs that have died out since Madagascar was colonized by human beings were as large or larger than the surviving species (Dewar 1984).

Species with poor dispersal and colonization ability. As with local endemics, species with narrow habitat requirements and species that can't disperse easily to new habitats are at high risk of extinction, even if their population is widespread. For example, in the face of a warming climate, some of the most threatened species will be those that can't disperse as fast as suitable habitat moves to higher elevations and latitudes.

Species with colonial nesting habits. Colonial nesting species are particularly susceptible to over-exploitation or the loss of breeding habitat. Several species of marine turtles, for example, have been threatened by the development of nesting beaches for tourism and other purposes. Similarly, the vulnerability of the extinct passenger pigeon (*Ectopistes migratorius*) to hunters was increased by the birds' tendency to nest in enormous colonies.

Migratory species. Migratory species depend upon suitable habitat in their summer and winter range *and* along the course of their migratory route. Thus, the potential for adverse effects of habitat changes on migrant populations is high. In North America,

Box 4. (cont.)

Kirtland's warbler *(Dendroica kirtlandi)*, Bachman's warbler *(Vermivora bachmanii)*, and the Whooping Crane *(Grus americana)* are endangered because habitat in both their breeding and wintering ranges has shrunk.

Species dependent on unreliable resources. These species' populations fluctuate greatly, so they face increased threats of extinction when their population is low. Populations of frugivores and nectarivores, for example, may be reduced significantly during years when fruit or nectar crops are low.

Species with little evolutionary experience with disturbances. In regions where human beings have a longstanding presence, the species most sensitive to human disturbance have already been lost and many of the remaining species have adapted to the additional disturbance. In contrast, species are extremely vulnerable where human disturbance has no historical precedent. Thus, the loss of the great auk, passenger pigeon, elephant birds, and moas, and the endangerment of the musk oxen *(Ovibos moschatus)* were all related to the introduction of a predator where none had existed before or to the introduction of a predator with a new means of hunting—say, the rifle—that the prey had not previously experienced.

Based on: Terborgh 1974, Pimm et al. 1988, and Orians and Kunin, in press.

---

amount lost reaches up to 94 percent in Bangladesh (IUCN/UNEP 1986a, b).

While less devastating than complete habitat loss, the fragmentation of habitat into patches can gradually wipe out species whose surviving individuals don't make up a *minimum viable population* (MVP) of the species. At numbers below the MVP, populations face an increased risk of extinction due to a combination of the following factors (Shaffer 1981, Soulé and Simberloff 1986, Soulé 1987):

• *Demographic Chance*—Chance variation in survival and reproduction. In wild populations, the chance variation in such demographic characteristics as birth and death rate and the sex ratio of offspring may threaten populations, particularly those smaller than 100 individuals (Lande 1988).

• *Environmental Chance and Natural Catastrophes*—Chance variation in weather, competitors, predators, and disease.

• *Genetic Chance*—Inbreeding, genetic drift, and loss of genetic variation. Inbreeding will impede species survival in populations that are rapidly reduced to less than about 50 breeding individuals. In populations of less than roughly 500 breeding individuals, genetic variation may be reduced, eventually diminishing the species' capacity to adapt to environmental change (Franklin 1980, Soulé 1980, Lande 1988).

• *Social Dysfunction*—Disruption of breeding, foraging, defense, thermoregulation, or other social behaviors in the absence of some critical number of individuals.

For small populations, sheer chance is a formidable force in extinction. Demographic chance sealed the fate of the dusky seaside sparrow *(Ammodramus maritimus nigrescens)* when the population fell to six individuals—all of which happened to be male (Avise and Nelson 1989). Environmental chance was partially responsible for the loss of the heath hen. Similarly, hurricane Gilbert in the Gulf of Mexico in 1988 devastated the only known breeding beach of the already endangered Kemp's ridley sea turtle *(Lepidochelys kempii)*. Genetic chance may be contributing to the endangerment of the golden lion tamarin *(Leontopithecus rosalia)* where inbreeding has apparently caused a high incidence of a genetic defect of the diaphragm (Soulé and Simberloff 1986). And social dysfunction has contributed to the decline of populations of wolves *(Canis lupis)*, since hunting success is reduced if pack size declines.

Populations below their minimum viable size will gradually disappear from islands unless individuals from outside recolonize it. Once the habitat is fragmented, species richness shrinks to the number capable of persisting in the reduced habitat area. One example of this so-called relaxation effect is the loss of species from Barro Colorado Island in Panama, which was created around 1914 when the Panama Canal was built and surrounding valleys were dammed. Of the approximately 200 land bird species known to have bred on Barro Colorado, 47 had disappeared by 1981: 21 of these species are thought to have been lost because the island was isolated from the formerly continuous habitat (Karr 1982); many of the remainder are believed to have been associated with early successional habitats that gradually disappeared when the island was protected from human disturbances.

For many wild populations, demographic factors appear more immediately threatening than genetic factors to the persistence of populations (Goodman 1987, Lande 1988). Habitat loss and fragmentation often seals a population's fate simply by reducing individual survival and breeding success before inbreeding or the lack of genetic variation begins to take its toll. When the U.S. Forest Service developed a management plan for the northern spotted owl *(Strix occidentalis caurina)* based on genetic considerations, a population of 500 individuals seemed sufficient to guarantee survival, but the habitat allowed to preserve these 500 individuals would have been so fragmented and sparsely distributed that the population would probably have become extinct as a result of increased juvenile mortality and decreased reproductive success (Dawson et al. 1987,

Simberloff 1987, Lande 1988). Consequently, estimates of minimum viable population size have been revised upward, and close attention is now being given by managers to the pattern of forest fragmentation.

The species least affected by the loss or degradation of habitat are those that are widespread or that disperse easily. In contrast, habitat loss or degradation is most threatening to highly specialized species, species with restricted ranges, and species with chronically low population sizes. *(See Box 4.)* Thus, many island and locally endemic plants and animals will be lost because of their susceptibility to habitat loss, and extinctions among top carnivores and large animals will be disproportionate because their population density is generally low. A region of relatively natural habitat may be large enough to maintain a small (7 kg) species such as a golden jackel *(Canis aureus)* with a home range of 7 square kilometers but far too small to maintain a population of tigers *(Panthera tigris;* 110–150 kg) with home range of 20 to 62 square kilometers.

Specific threatened habitats of particular value both for the species diversity that they contain and for their provision of other services to humanity include temperate old-growth forests, tropical forests, wetlands, and coral reefs.

Temperate Forests. Conserving the biodiversity of temperate forests requires the maintenance of all successional stages of the forest (Franklin 1988). Unlike forests in early successional stages, old-growth forests provide a unique physical structure and consequently a unique habitat for plants and animals. In the Tongass National Forest, a temperate coniferous rain forest in Alaska, old growth provides critical winter habitat for the Sitka black-tailed deer *(Odocoileus hemionus sitkensis)* and important breeding habitat for the Canada Goose *(Branta canadensis fulva)*. Ecological processes also differ among successional stages. For example, nutrient losses from old-growth watersheds in the Pacific Northwest of the United States are lower than the losses from early successional stages (Franklin 1988).

Excessive logging, conversion of forest to city and farm, and pollution significantly threaten the biodiversity of temperate forests. In particular, old-growth forests have been eliminated throughout much of their native range, and some of the last regions with extensive old growth are now slated for timber harvest. In northern California, Oregon, and Washington, only 15 to 30 percent of the 19 million hectares of old growth that existed in the 19th century remains (Booth 1989).

Tropical Forests. The extent of closed tropical forest worldwide has been reduced by 23 percent from the estimated historic area, and 15 percent of the remaining forest is in use for timber production. *(See Table 18.)* Each year, at least 7.4 million hectares of the remaining natural forest is deforested or logged. Given the vast tropical forests remaining in the Amazon and Zaire, complete loss does not seem imminent, but both specific types of tropical forest and forested areas in certain regions are on the verge of disappearing. Brazil's Atlantic forest has been reduced to less than 5 percent of its original 1 million square kilometers (Mori et al. 1983, Myers 1988). In the state of Minas Gerais, less than 7 percent of the original forested area has any cover remaining and virtually all undisturbed forest is gone (Fonseca 1985).

In parts of the tropics, the most threatened forest type is the tropical dry forest. This forest appears where mean annual rainfall ranges from 250 to 2000 mm. Many such regions have a pronounced dry season, so a high proportion of tree species are deciduous. Dry forests have been extensively cleared because they are found in climates well-suited to agriculture and grazing. In five Central American countries, the average human population density in the dry forest biome is eight times that in the rain forest (Murphy and Lugo 1986). The Pacific coast of central America contained 550,000 square kilometers of dry forest at the time of

the arrival of Europeans; now, less than 2 percent is intact (Janzen 1988a).

Wetlands. The extent of both coastal (brackish or saline) and inland (freshwater) wetlands has been significantly reduced by human activities. The United States has lost 54 percent of its wetlands since colonial times, Brittany has lost 40 percent in the past 20 years, and almost 20 percent of the internationally important wetlands in Latin America are threatened by drainage related to development activities (Maltby 1988).

Mangroves are one of the world's most productive ecosystems: many produce far more charcoal, poles, and firewood per hectare than inland forest does and all provide a barrier against coastal erosion.

One of the most threatened wetland types, particularly in light of its tremendous value to humanity, is the mangrove ecosystem. A type of forest found in tropical coastal wetlands, mangroves grow on muddy, saline, and often anaerobic sediments in areas with freshwater runoff. Mangroves cover some 240,000 square kilometers of intertidal and riverine ecosystems, reaching their greatest extent along the coasts of south and south-east Asia, Africa, and South America (WRI/IIED 1986). These forests support a moderate diversity of terrestrial animals and provide shelter and breeding places for many commercial species of prawns and fish. In Fiji, roughly half of the catch of commercial and artisanal fisheries is of species that depend on mangrove areas during at least one critical stage of their life cycle. In eastern Australia, 67 percent of the commercial catch is composed of species dependent on mangrove communities (WRI/IIED 1986). Mangroves are

49

one of the world's most productive ecosystems: many produce far more charcoal, poles, and firewood per hectare than inland forest does and all provide a barrier against coastal erosion.

The rate of loss of mangrove forests is extremely high. Thailand has lost 22 percent of its mangroves since 1961 and 8 percent between 1975 and 1979 alone (FAO 1985a). Virtually none of the remaining Thai mangrove forests are undisturbed (J. McNeely, IUCN, personal communication, March 1989). The Philippines have lost half of their mangroves since the turn of the century and 24,000 hectares annually between 1967 and 1975 (FAO 1982, 1985a). In peninsular Malaysia, 20 percent of mangrove area has been converted to other uses in the past 20 years.

Despite the value of mangroves to the fishing industry in both Southeast Asia and Latin America, one of the principal causes of the loss of this key resource has been aquaculture development. Many mangrove estuaries are converted to brackish ponds for the culture of shrimp and prawns. An estimated 1.2 million hectares of mangrove in the Indo-Pacific region is currently used for aquaculture, almost 15 percent of the total (ESCAP 1985). Other threats to mangroves include the conversion to rice fields (i.e., Senegal, Gambia, and Sierra Leone) and coconut plantations, and over-harvesting for timber and fuelwood.

Coral Reefs. Coral reefs and their associated communities cover an estimated 600,000 square kilometers, mostly between the latitudes 30°N and 30°S (Smith 1978). *(See Box 2.)* Coral reefs extend to depths of 30 meters and cover 15 percent of the world's coastline. Fish production on these reefs and on the adjacent continental shelf could amount to nearly 10 percent of global fisheries production if fully exploited (Smith 1978). Locally, coral reefs are even more important as a food source. Throughout southeast Asia, coral reef fisheries provide 10 to 25 percent of the protein available to people living along the coastlines

(McManus 1988). Coral reefs also protect coastal areas from erosion. In the case of coral atolls, coral provides the foundation of the island itself. In the Indian Ocean, 77 percent of isolated islands and island archipelagoes are built exclusively of reef depositions (Salm and Clark 1984).

Unfortunately, sediment pollution resulting from erosion in upland watersheds, coral mining, thermal and chemical pollution, and the use of dynamite for fishing have seriously degraded the productivity of reef ecosystems throughout the world. In the Philippines, some 40 percent of all coral reef is in a poor state (Meith and Helmer 1983).

In response to the threats to reef communities, some 101 countries have some form of coral reef protected areas, but many of these sites have little or no financing for management or enforcement (Salm and Clark 1984). For example, 114 marine protected areas have been established in the Caribbean, but 27 percent of them have no management or enforcement capacity (OAS 1988).

Over-exploitation

Some 37 percent of all endangered, vulnerable, and rare species of vertebrates are threatened by over-exploitation (Prescott-Allen and Prescott-Allen 1978). Many fur-bearing animals—including chinchilla (*Chinchilla spp.*), vicuña (*Vicugna vicugna*), giant otter (*Pteronura brasiliensis*), many species of cats, and some species of monkeys—have declined to very low population sizes because their pelts are prized. Valuable timber species including populations of the West Indies mahogany (*Swietenia mahogoni*) in the Bahamas and the Caoba "mahogany" (*Persea theobromifolia*) of Ecuador have been severely depleted, and the Lebanese cedar (*Cedrus libani*)—which once covered 500,000 ha of Lebanon—has been reduced to a few scattered remnants of forest (Chaney and Basbous 1978, Oldfield 1984). On Mauritius and Reunion islands, an excellent hard timber species, bois de prune blanc (*Drypetes caustica*), was reduced almost to extinction; only 12 trees remain (Prance, in press). Several species of sea turtles are seriously threatened by over-exploitation (for eggs, meat, shells, and oil) and the loss of nesting sites. African elephant (*Loxodonta africana*) populations in Africa declined by 31 percent (from 1.2 million to 764,000) between 1981 and 1987, largely because of the demand for ivory (IUCN 1987).

Uncontrolled commercial harvest of a species tends to reduce its population to "commercial extinction." As the population size declines, the cost of capture increases until further exploitation becomes too expensive. If a species becomes commercially extinct at a relatively high population size, the species may be in no danger of actual extinction. But if the commercial value of the species is extremely high or the species is relatively easy to find and capture, then overharvesting could lead to extinction.

Such is the case with the rhinoceros and potentially with the elephant as well. The retail price of rhinoceros horn in the Middle East and Asia ranges from $800 to $13,500 per pound (Sheeline 1987). Since the average rhino horn weighs roughly 3.3 pounds, each has a retail value of between $2,600 and $44,500, depending on quality and the market. These high values create extremely high pressures on the remaining rhinoceros population. In 1986, it was estimated that fewer than 50 individuals of the northern sub-species of the white rhinoceros (*Cerathotherium simum cottoni*) remained alive in the wild. The last five white rhinos in Kenya's National Parks were killed after heavily armed poachers overwhelmed armed rangers guarding the animals in Meru National Park (Ransdell 1989). The population of the black rhinoceros (*Diceros bicornis*) of East Africa has dropped precipitously since 1970 (Hillman-Smith et al. 1986). Asian species of rhinos are even more threatened. Only 700 individuals of the Sumatran rhino (*Dicerorhinus sumatrensis*) remain, and only 55 of the Javan rhinos (*Rhinoceros sondaicus*) remain. For all five species of rhinoceros in the world, the population has declined by 84 percent since 1970 (Sheeline 1987).

Over-exploitation is a more selective threat to species survival than is habitat loss. Whereas habitat loss threatens a wide range of taxa from plants and invertebrates to vertebrates, over-exploitation primarily threatens vertebrates and certain taxa of plants and insects. More specifically, carnivores, ungulates, primates, sea turtles, showy tropical birds, and timber species have been overharvested. Many species of butterflies and orchids have been overharvested for commercial interests too, and rare plants have been threatened by collectors. Predator-control efforts have also significantly reduced the population sizes of many vertebrates, including sea lions, birds of prey, foxes, wolves, various large cats, and bears. Because such top-predators often play important roles in structuring communities, predator-control efforts are likely to deeply affect the populations of other species in the ecosystems too. *(See Appendix 1.)*

Over-exploitation often acts synergistically with habitat loss or fragmentation to increase the threat of species loss. Road construction, for example, not only fragments the existing habitat but also allows increased access for hunting, collecting, and timber harvesting.

Species Introductions

Introduced species threaten 19 percent of all endangered, vulnerable, and rare species of vertebrates (Prescott-Allen and Prescott-Allen 1978). They also affect plants and invertebrates, particularly on islands. Some 86 of Hawaii's 909 exotic plant species that are reproducing in the wild are considered serious threats to native ecosystems (Vitousek 1988, W. Wagner, Smithsonian, personal communication, April 1989).

Exotic species can threaten native flora and fauna directly by predation or competition or indirectly by altering the natural habitat. On the Galápagos Islands, for example, black rats

(*Rattus rattus*) have reduced populations of the giant tortoise (*Geochelone elephantopus*) and the dark-rumped petrel (*Pterodroma phaeopygia*) by preying on eggs (Hoeck 1984), and have wiped out some rodent species on the islands. Other exotic species on the Galápagos, such as cattle, goats, and feral pigs, have changed the vegetation on some islands from highland forest into pampa-like grasslands and have destroyed stands of cactus (Hoeck 1984). Several species of trees on islands with many goats can't propagate because the goats eat the saplings, and habitat changes caused by the goats have reduced populations of some races of tortoises and land iguanas.

Pollution

Although pollution is not one of the primary causes of species extinctions, pollution's harm to animal populations first catapulted the problem of endangered species into high public visibility in 1962, when Rachel Carson brought the dramatic effects of DDT and other pesticides on wildlife populations to the public's attention in *Silent Spring*. More important, by the middle of the next century, pollution may rival habitat loss as the most important threat to species if rapid global warming occurs.

By the middle of the next century, pollution may rival habitat loss as the most important threat to species if rapid global warming occurs.

Both water and air pollution stress ecosystems and reduce populations of sensitive species. For example, air pollution has caused declines in several species of eastern pines in the United States and has been linked to forest diebacks in Europe (Bormann 1982, MacKenzie and El-Ashry 1988). After exposure to air pollution and acid rain, forests can experience

outbreaks of insects and pests, population declines of sensitive species, and reduced productivity. As a result, their capacity to provide immediate services to humanity is impaired and the likelihood of species extinctions increased. Acid rain also harms freshwater organisms sensitive to changes in water acidity. With increasing industrialization and continued economic growth adding ever greater loads of pollutants to the land, air, and water (Speth 1988), environmental pollution will become an increasing threat to biological diversity in coming decades even apart from the effects of climate change.

Global Warming

How much pollution influences species extinction may change dramatically in the coming decades. During the next century, global average surface temperatures are expected to increase by 2° to 6°C over current levels unless greenhouse gas emissions are reduced immediately (Schneider 1989). Global warming has the potential to significantly affect the world's biological diversity: it could cause changes in the population sizes and distributions of species, modify the species composition of habitats and ecosystems, alter the geographical extent of various habitats and ecosystems, and increase the rate of species extinction.

In general, if temperatures rise, regions of suitable climate for given species will shift toward the poles and toward higher elevations. In practice, however, the pattern of change will be far more complex. In the wake of climate change, patterns of precipitation, evaporation, and wind and the frequency of storms, fire, and other disturbances will all influence the location and suitability of habitats. Ecosystems are not cohesive assemblages of species that will "migrate" as a unit in response to climate change. Instead, the species in plant and animal communities will probably dissociate as a result of differences in thermal tolerances, habitat requirements, and dispersal capacity and then reassemble in new configurations under the new climate regime. How much green house warming affects species diversity and the composition, distribution, and extent of habitats and ecosystems will depend on the magnitude and rate of climate change and on the species' differential capacity to respond through dispersal, colonization, and (particularly in the case of corals) growth.

In the temperate zone, both the magnitude of forecasted warming and the rapid rate of temperature change—somewhere between 10 and 60 times higher than the average rate of temperature rise after the last Ice Age (Schneider 1989)—suggest that impacts on biodiversity may be severe. Temperate deciduous forests could experience range shifts of 500 to 1,000 km if temperatures rise by 2° to 6°C (Davis and Zabinski, in press). The rate of the poleward shift in suitable habitat may exceed natural dispersal rates of many species by more than an order of magnitude. During the post-Pleistocene warming, most North American trees migrated at rates of only 10 to 40 km per century (Davis and Zabinski, in press). In tropical regions, differences in temperature and precipitation may not be as large, but potential changes in the seasonal timing of rainfall or in the frequency or intensity of such disturbances as hurricanes or fires could have equally serious effects on the species composition of tropical ecosystems.

Although global climate models are not sufficiently accurate to allow detailed predictions of regional patterns of temperature and precipitation, one consequence of global warming can be predicted with greater accuracy: sea level rise. When sea level rises slowly, coastal wetlands that are not blocked by coastal development follow the rise by moving landward and vertically accumulating peat. (See Figure 8.) If sea level rises faster than wetland can build up, however, substantial area may be lost (Smith and Tirpak 1988). It is estimated that a one-meter rise in sea level over the next century—on average, what scientists are now predicting—would destroy 25 to 80 percent of United States coastal wetlands (Smith and Tirpak 1988). This figure could go higher if

wetlands' movement landward is blocked by coastal development, bulkheads, and levees.

Both sea level rise and changing water temperatures may also threaten coral reef ecosystems (Buddemeier and Smith 1988). Current rates of sea level rise are on the order of 3 mm per year (Ray et al., in press), well within the maximum potential of coral reefs to grow upward, estimated to be 10 mm per year (Grigg and Epp 1989). However, a sea level rise of 0.5 to 1.5 meters during the next century would entail an average yearly rise of 5 to 15 mm and could drown coral reefs and threaten the tremendous diversity of their associated species. The species composition of reefs could also change because certain kinds of coral—branching corals—are capable of more rapid growth and may outcompete other corals. Elevated water temperatures could further threaten coral reef ecosystems. Abnormally high sea-water temperatures in 1982 and 1983, for instance, may have contributed to coral "bleaching"—the loss of the mutualistic protozoans living with the coral—and death. At any rate, coral reefs in the eastern Pacific had lost an estimated 70 to 95 percent of their living coral to depths of some 18 meters by the end of 1983 (UNEP/IUCN 1988).

Species' capacity to shift their ranges in response to climate change will be hindered by human land-use practices that have fragmented existing habitats. Much of the world's species diversity is contained in relatively small habitat patches—parks, national forests, and other protected areas—surrounded by agricultural or urban landscapes that many species can't cross. And even when species have the capacity to disperse to new sites, suitable habitat may not be available. In the case of global warming, alpine habitats will disappear from many mountains and the use of levees, dikes, and sea-walls to protect coastal development against sea-level rises will decrease coastal wetland habitat.

Human beings may pay a price in adapting to climate change that exceeds the costs of

Figure 8: Change in Wetland Area as Sea Rises

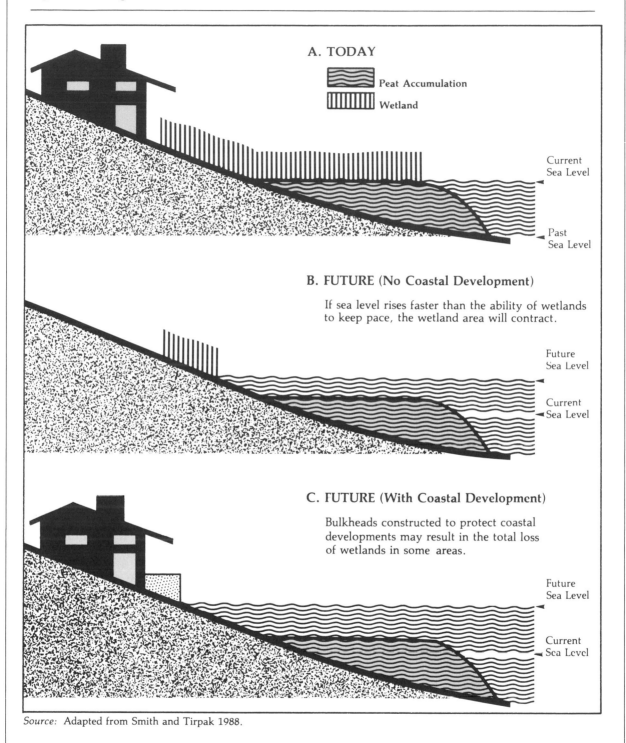

A. TODAY

[legend] Peat Accumulation

[legend] Wetland

Current
Sea Level

Past
Sea Level

B. FUTURE (No Coastal Development)

If sea level rises faster than the ability of wetlands
to keep pace, the wetland area will contract.

Future
Sea Level

Current
Sea Level

C. FUTURE (With Coastal Development)

Bulkheads constructed to protect coastal
developments may result in the total loss
of wetlands in some areas.

Future
Sea Level

Current
Sea Level

Source: Adapted from Smith and Tirpak 1988.

losses of species and their associated ecosystem services. For instance, as species shift ranges, the regional prevalence of certain human and agricultural diseases could increase. In tropical regions, the range of the tsetse fly (*Glossina* spp.) and the parasite that it transmits (*Trypanosomiasis* spp., which causes sleeping sickness in humans and the disease, nagana, in domestic animals) could shift (Dobson and Carper, in press). If it does, harm to human populations currently outside of the tsetse fly range can be expected, and some areas previously off-limits to human exploitation would be settled, with commensurate losses of habitats and many of their associated species.

Cumulative Effects

Human disturbances of ecosystems rarely involve just a single impact. Far more common is a pattern of repeated or simultaneous disturbances. For example, the fragmentation of a forest causes species extinctions, changes hydrological cycles, changes population sizes, and alters migratory patterns; at the same time, the forest may be independently disturbed by acid rain and climate change. Similarly, over-exploitation of one fish may be compounded by over-exploitation of another in the ecosystem, marine pollution, and loss of spawning habitat.

The cumulative effect of various disturbances can seriously threaten a region's biodiversity. Even if the disturbances have only additive effects, small changes can add up to serious impacts (NRC 1986). Ehrlich and Ehrlich (1981) provide an apt analogy for this problem: a man prying rivets out of the wing of an airliner so that he can sell the rivets sees no reason to worry about the consequences of his actions since he has already removed numerous rivets from the wing with no apparent ill effect.

The loss of wetland habitat dramatically illustrates the potential cumulative impacts of relatively small environmental disturbances. The loss of one coastal wetland would have little effect on fish populations dependent on wetlands for one portion of their life cycle; but if other wetlands in the region are also lost, the harm could be great. Between 1950 and 1970, Connecticut and Massachusetts lost about half of their coastal wetlands. Hundreds of local decisions were to blame, and no single decision reflected the gravity of the cumulative effect. Odum (1982), following economist Alfred Kahn, refers to this process as the "tyranny of small decisions."

Even more troubling, some disturbances may be synergistic rather than merely additive. Such interactions may have larger or more damaging effects on ecosystems than additive impacts do (NRC 1986). For example, habitat fragmentation alone in a constant climatic regime, or climate change in a region with an unfragmented habitat, will have far less impact on species diversity than will the two factors combined. Species that would have otherwise been able to migrate to suitable climatic zones in a continuous habitat may be lost if physical barriers fragment the habitat. For instance, the heath hen described earlier was wiped out by a synergistic effect of habitat loss, over-exploitation, and unusual weather.

Resource management and land-use planning must take account of cumulative effects. Indeed, without wide-ranging strategic decisions, seemingly disconnected small actions could cumulatively degrade important resources. In the worst case, rainforests, wetlands, coral reefs, and other vast resources could be nickelled and dimed to death.

V. What's Happening to Agricultural Genetic Diversity?

The management of genetic diversity is critical to the maintenance of all wild and captive populations, but its value is particularly apparent in agriculture where it has for millennia been used to enhance productivity. Tremendous advances have been made in the past 20 years to secure agricultural genetic diversity, but much more work remains to be done. Currently, the conservation of genetic diversity and its utilization in the development and patenting of new life forms is hotly debated in North-South dialogues and by resource managers. *(See Plucknett et al. 1987; Kloppenburg and Kleinman 1987; Kloppenburg 1988a, b; Juma 1989.)* And no consensus can be expected until the distribution and status of the genetic diversity of economically important species is widely understood.

Uses of Genetic Diversity

In agriculture and forestry, genetic diversity can enhance production. Several varieties can be planted in the same field to minimize crop failure, and new varieties can be bred to maximize production or adapt to adverse or changing conditions. The Massa of northern Cameroon (who cultivate five varieties of pearl millet), the Ifugao of the island of Luzon in the Philippines (who identify more than 200 varieties of sweet potato by name), and Andean farmers (who cultivate thousands of clones of potatoes, more than 1,000 of which have names) all use highly diversified farming systems—the first approach (Clawson 1985).

The second approach is also widely used. In the United States from 1930 to 1980, plant breeders' use of genetic diversity accounted for at least one-half of a doubling in yields of rice, barley, soybeans, wheat, cotton, and sugarcane; a threefold increase in tomato yields; and a fourfold increase in yields of corn, sorghum, and potato (OTA 1987).

In general, the easy gains in plant breeding come first; then, the crop's response is less dramatic and increasingly greater research and breeding are needed to sustain increased yields.

As important as genetic diversity is to increasing yields, it is at least as important in maintaining existing productivity. For example, crop yields can be increased by introducing genetic resistance to certain insect pests, but since natural selection often helps insects quickly overcome this resistance, new genetic resistance has to be periodically introduced into the crop just to sustain the higher productivity. Pesticides are also overcome by evolution, so another important agricultural use of genetic diversity has been to offset productivity losses from pesticide resistance. Indeed, the record shows that pesticides only temporarily conquer pests. Over 400 species of pests now resist one or more pesticides (May 1985), and the proportion of U.S. crops lost to insects has approximately doubled—to 13 percent—since the 1940s, even though pesticide use has increased (Plucknett and Smith 1986). Even more significantly, as crop yields increase, so must efforts to sustain the gains (Plucknett and Smith 1986). In general, the easy gains come first.

Until the advent of genetic engineering, the spectrum of genetic resources available for plant breeding ranged from other varieties of the crop to wild relatives of the species. Wild relatives of crops have contributed significantly to agriculture, particularly in disease resistance (Prescott-Allen and Prescott-Allen 1983). Thanks to wild wheats, domesticated wheat now resists fungal diseases, drought, winter cold, and heat. Rice gets its resistance to two of Asia's four main rice diseases from a single sample of wild rice from central India (Prescott-Allen and Prescott-Allen 1983).

The spectrum of genetic resources available to breeders is now expanding to still more distantly related species, thanks to genetic engineering techniques. However, much more work remains to be done before the potentials of these techniques are realized (Plucknett et al. 1987). Moreover, most new genetic engineering techniques will at least initially involve single-gene modifications of species, and in many cases such modifications are less useful than the multiple-gene changes that result from traditional breeding programs. A transfer of a single gene to a crop plant may confer resistance to a certain pest, but that resistance is often relatively quickly overcome by the pest. In only one species—tobacco—has single-gene resistance been useful for extended periods of time (Prescott-Allen and Prescott-Allen 1986).

Genetic diversity has also been of significance in breeding programs for species other than edible plants. It is becoming increasingly important in forestry and fisheries, and the use of genetic resources in livestock breeding has markedly increased yields. The average milk yield of cows in the United States has doubled over the past 30 years, and genetic improvement accounts for more than 25 percent of this gain in at least one breed (Prescott-Allen and Prescott-Allen 1983).

Still, for several reasons, genetic diversity has been less useful in livestock breeding than in crop breeding (Frankel and Soulé 1981). First, whereas one major use of the genetic diversity of crops has been in the development of strains resistant to specific pests and diseases, livestock husbandry has relied largely on vaccines since animals (unlike plants) can develop immunity to disease. Second, maintaining livestock germplasm is tougher logistically than maintaining the genetic material of plants: since domesticated animals do not go through dormant stages comparable to the seed stage of plants, long-term storage is a problem. Finally, many of the closest relatives of domesticated animals are extinct, endangered, or rare, and thus unavailable for breeding (Prescott-Allen and Prescott-Allen 1983).

For different reasons, genetic improvement of forest species has also received less attention than crop improvement. Until recently, most timber harvested has been wild, so breeding programs seemed unnecessary. In addition, because trees are so long-lived, the rate of genetic improvement of tree species is quite slow. Tests and measurements of growth characteristics have been made for some 500 species (primarily conifers) over the years, but less than 40 tree species are being bred (G. Namkoong, North Carolina State University, personal communication, May 1989). Yet, impressive gains have been made with these species. In intensive breeding programs, a 15 to 25 percent gain in productivity per generation has been attained for trees growing on high-quality sites without inputs of fertilizer, water, or pesticides (G. Namkoong, personal communication, May 1989).

Most of the fish that humanity eats or converts to livestock feed are wild, so breeding has not been widely utilized in fisheries to enhance yields either. An exception is aquaculture. In one case, the domestic carp (*Cyprinus carpio*) was bred with a wild carp in the Soviet Union to enhance the cold resistance of the domestic species and allow a range extension to the north (Prescott-Allen and Prescott-Allen 1983).

The development of high-yielding varieties of crops has greatly benefitted some segments of society, but some risks attend use of these new varieties. With the advent of modern plant breeding techniques, a trend toward genetically more uniform agriculture has developed in areas suited to the high-yielding, high-input modern varieties. Whereas traditional mixed farming systems produce modest but reliable yields, planting a single modern crop variety over a large area can result in high yields but the crop may be extremely vulnerable to pests, disease, and severe weather. In 1970, for instance, the U.S. corn crop suffered a 15-percent reduction in yield and losses worth roughly $1 billion when a leaf fungus (*Helminthosporium maydis*) spread rapidly through the genetically uniform crop (Tatum 1971). Similarly, the Irish potato famine in 1846, the loss of a large portion of the Soviet wheat crop to cold weather in 1972, and the citrus canker outbreak in Florida in 1984 all stemmed from reductions in genetic diversity (Plucknett et al. 1987).

To stabilize production, breeders use one of several tactics to maintain a genetically diverse crop array. Typically, varieties are replaced with higher-yielding relatives after four to ten years even if they still resist disease or pests. In effect, the spatial diversity of traditional agriculture is replaced with a temporal diversity created by a continuous supply of new cultivars. In the United States, the average lifetime of a cultivar of cotton, soybean, wheat, maize, oats, or sorghum is between 5 and 9 years (Plucknett and Smith 1986). An example of a 10- to 12-year cycle of varieties of sugarcane is shown in Figure 9.

Newer strategies for stabilizing production involve the use of varietal blends (a mix of strains sharing similar traits but based on different parents) or multilines (varieties containing several different sources of resistance). In each case, the crop represents a genetically diverse array that can better withstand disease and pests (Plucknett et al. 1987). Despite these efforts, genetic uniformity still places some crops at risk of disease outbreaks and in some regions that risk is considerable. Some 62 percent of rice varieties in Bangladesh, 74 percent in Indonesia, and 75 percent in Sri Lanka are derived from one maternal parent (Hargrove et al. 1988).

In Situ Conservation

Maintenance of plant and animal genetic material in the wild (*in situ*) and maintenance of wild or domesticated material in gardens, orchards, seed collections, or laboratories (*ex situ*) are both essential aspects of managing the human use of genetic diversity. *In situ* conservation maintains not only a variety's genetic diversity but also the evolutionary interactions that allow it to adapt continually to shifting environmental conditions, such as changes in pest populations or climate. In agriculture, *in situ* conservation is also the best way to maintain knowledge of the farming systems in which the varieties have evolved.

One innovative technique that has been suggested for maintaining both genetic diversity and knowledge of farming systems where traditional varieties are being lost involves the use of village-level "landrace custodians." At a limited number of key sites (perhaps between 100 and 500) in areas where diseases, pests, and pathogens strongly influence the evolution of local crop varieties, individuals could be paid to grow a sample of the endangered native landraces (Altieri et al. 1987; Wilkes, in press[a]). These farms would backstop *ex situ* conservation efforts, maintain the potential for further evolution in important landraces, preserve knowledge of traditional farming systems, and provide regional education on the importance of biodiversity conservation. If it seems economically inefficient to subsidize the

AUGUSTANA UNIVERSITY COLLEGE
LIBRARY

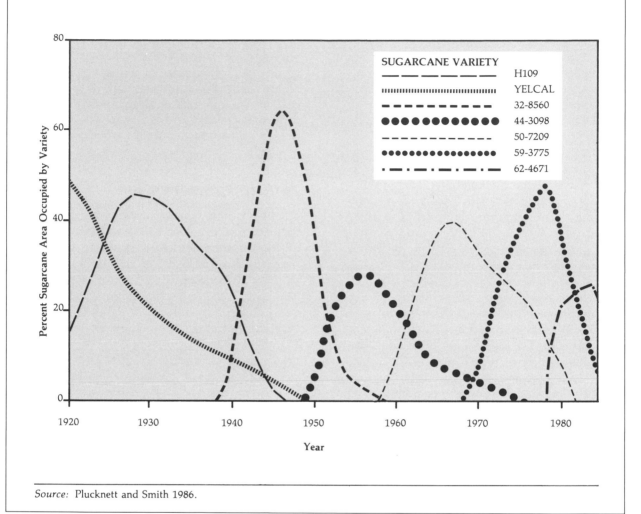

Figure 9: Pattern of Replacement of Varieties of Sugarcane in Hawaii, 1920-1984

SUGARCANE VARIETY

— — — — — —	H109
׀׀׀׀׀׀׀׀׀׀׀׀׀׀׀׀׀׀׀׀	YELCAL
– – – – – – –	32-8560
●●●●●●●●●●	44-3098
- - - - - - -	50-7209
••••••••••••••	59-3775
–·–·–·–·–·–	62-4671

Source: Plucknett and Smith 1986.

in situ maintenance of traditional landraces, consider how often their loss is hastened by subsidies for such agricultural inputs as fertilizers and pesticides. For instance, in 1985, Indonesia had an 85-percent subsidy on pesticides, at an annual cost of $120 million (Hansen 1987).

The greatest untapped potential for *in situ* germplasm conservation resides in protected areas set aside to conserve species that cannot be preserved *ex situ* and wild crop relatives. A number of species (an estimated one-third of plants in lowland tropical forests; P. Raven, personal communication, April 1989) possess recalcitrant seeds—seeds unsuited to long-term storage in seed banks. With greater research, better storage techniques may eventually be developed, but for now *in situ* conservation represents one of the few conservation techniques available. In addition, *in situ* conservation of wild crop relatives maintains not only the target species, but also a host of other species that share the same habitat.

The International Board for Plant Genetic Resources (IBPGR) has identified numerous species whose wild relatives are prime candidates for *in situ* conservation. These include relatives of groundnut, oil palm, banana, rubber, coffee, cocoa, members of the onion family, citrus fruits, mango, cherries, apples, pears, and many forage species. Some pilot efforts are now under way to develop such *in situ* conservation programs. A gene sanctuary has been established in India for *Citrus* species, two reserves have been established in Zambia to conserve Zambesi teak *(Baikiaea plurijuga)*, and one has been set up in Canada for jack pine *(Pinus banksiana)* (FAO 1985b). The existing 4,500 protected areas monitored by the World Conservation Monitoring Center can be expected to harbor many wild relatives, though no actual field inventory has been conducted.

Little effort has been directed at conserving the genetic diversity of forestry species. A number of populations of the two dozen species that have been intensively bred survive in various types of protected areas, so it seems unlikely that the genetic base of the species will shrink. However, several hundred wild or semi-domesticated tree species of economic value throughout the world are at risk of being lost in all or part of their ranges. In a review in 1986, the FAO identified 86 such threatened species (FAO 1986), and this list has lengthened since. Only about half of the populations threatened in 1989 are represented within protected areas (G. Namkoong, personal communication, May 1989).

Several hundred wild or semi-domesticated tree species of economic value throughout the world are at risk of being lost in all or part of their ranges.

Wild forest species are particularly susceptible to genetic erosion because of the practice of high grading, whereby the best trees are removed each harvest cycle and the less useful trees become the breeding stock. This type of genetic erosion has reduced a number of once vigorous timber-producing forests in Japan, Korea, Turkey, and the Himalayas to stands of stunted and malformed trees (C. Miller 1987). The best defense against the erosion of forest genetic diversity is a combination of *in situ* conservation, *ex situ* storage, and the use of living collections of plants (known as *ex situ* field gene banks). How long the seeds of many tree species will last in storage has not yet been tested; particularly among tropical species, long-term storage may be difficult and field gene banks necessary. In addition, even species whose seed can be stored in seedbanks will have to be regenerated periodically; given the long generation time of forest species, continual maintenance in field gene banks will be needed (Palmberg 1984).

The best defense against the erosion of forest genetic diversity is a mixture of **in situ** *conservation,* **ex situ** *storage, and the use of living collections of plants.*

Ex Situ Conservation

Genetic diversity can be preserved *ex situ* through various techniques. In plants, the seeds of many species with so-called "orthodox seeds" can be stored in dry, low-temperature, vacuum containers (cryogenic storage). For some of these species, storage at extremely low temperatures, below −130°C, may extend the storage life to more than a century. In contrast, species with recalcitrant seeds can be maintained only *in situ* or in field collections, botanic gardens, and arboretums. Many species with recalcitrant seeds, particularly species that can be grown from cuttings, such as banana or taro, can be maintained by growing plant tissue or plantlets under specific conditions in glass or plastic vessels (*in vitro* culture). Recent work on cryogenic storage of *in vitro* cultures may further enhance this technique's potential. Cryopreservation also holds considerable promise for the long-term maintenance of livestock semen and embryos. Semen from at least 200 different species has been frozen, as have embryos from at least ten mammalian species (OTA 1987).

The chief benefit of *ex situ* preservation is in providing breeders with ready access to a wide range of genetic materials already screened for useful traits. *Ex situ* preservation may also represent a last resort for many species and varieties that would otherwise die out as their habitat is destroyed or modern varieties of plants or animals take their place. *In situ* conservation is often less expensive than *ex situ* techniques (Brush 1989), it insures against loss of *ex situ* collections, and it allows the continuing evolution of the crop varieties. *In situ* conservation also preserves knowledge of the farming systems with which local varieties evolved. Thus, the *ex situ* and *in situ* techniques complement each other and must be used together.

Ex situ *storage should be considered preservation rather than conservation.*

While some *ex situ* technologies—such as seed storage—are now extensively utilized, many problems with their use persist. Even in standard seed banks, the long-term integrity of the germplasm remains in question. Inadvertent selection or unintentional crossing with other varieties may occur, and plants stored *in vitro* mutate at relatively high rates. Perhaps most significantly, under any *ex situ* storage conditions, the evolution of the species is "frozen" so no further adaptation to pests or environmental changes can take place. For this reason, *ex situ* storage should be considered preservation rather than conservation.

The initial spread of new high-yielding varieties during the Green Revolution was dramatic. For example, modern varieties were adopted on 40 percent of Asia's rice farms within 15 years of their release, and in certain countries, such as Indonesia, Sri Lanka, and the Philippines, the comparable figure is more than 80 percent (Brush 1989). These threats to local crop varieties led the Consultative Group on International Agricultural Research (CGIAR) to established IBPGR in 1974 to stimulate and coordinate efforts to preserve the remaining genetic diversity of crops for future use in breeding programs. Since its inception, IBPGR has catalyzed improvements in the status of crop genetic resources by establishing conservation priorities, providing funds for exploration, and helping coordinate the seed-collection activities of the 13 CGIAR international agricultural research centers and 227 national seed banks in 99 countries.

Thanks to these activities, the status of crop genetic diversity has improved significantly in the last two decades. Today, reasonably comprehensive collections of the genetic diversity of most crops of importance in the temperate zone have been established, and the coverage of tropical crops is improving. *(See Table 13.)* For many of the major staple foods—including wheat, corn, oats, and potatoes—more than 90 percent of the landraces have now been preserved in *ex situ* collections (Plucknett et al. 1987). For many other globally important crops, the major part of the work involved in preserving landraces will soon be completed (Plucknett et al. 1987).

Despite such heartening progress, many obstacles remain in the quest to provide a secure source of germplasm. One is a lack of information. For example, for nearly half of the two million "accessions" (collections of seed from a specific locality) to gene banks worldwide, the plant's characteristics and the location where it was collected aren't recorded

(Wilkes, in press[b]). Then too, the high cost of *ex situ* collections, particularly when seed is stored at very low temperatures, may force some seed banks to cut back or shut down. At a minimum, high storage costs mean that funds for describing the germplasm present in the banks—a necessity for making the germplasm useful to plant breeders—are scarce.

But the most serious problem associated with *ex situ* collections involves gaps in coverage of important species, particularly those of significant value in tropical countries. The most worrisome gaps are in the coverage of species of regional importance, species with recalcitrant seeds, wild species, and livestock.

Because many important subsistence crops in developing countries are not widely traded in world markets and because storing germplasm of many tropical species is difficult, many regionally important crops are poorly represented in germplasm banks.

Crops of Regional Importance. Germplasm collection initially focused on food crops of greatest value in world commerce. Because many of the most important subsistence crops in developing countries are not widely traded in world markets and because many tropical species possess recalcitrant seeds, many regionally important crops are poorly represented in germplasm banks. Cassava and sweet potatoes are the fourth and fifth most important crops in the developing world by weight; yet, fewer than half of the landraces and fewer than 5 percent of the wild relatives are represented in *ex situ* collections. In addition, many species with either industrial or medicinal value are underrepresented in seedbanks. In 1987 the Food and Agriculture Organization of the U.N. noted that:

"[R]oots and tubers provide an important staple food for more than 1,000 million people in the developing world. Yet, unlike cereals, roots and tubers generally receive low priority—or no priority at all—in the agricultural plans of developing nations.... [The production of cassava, yams, sweet potato and taro] has suffered from years of neglect. Little of the research that has been done on increasing yields or improving storage characteristics has yet been applied. Where harvests have improved, it is usually because the area planted has been increased." (FAO 1987b)

The international seedbanks' priorities are beginning to shift to respond to this need to store germplasm from important species that aren't globally traded. In 1980, IBPGR initiated a major effort to conserve vegetable crop germplasm, for instance. Still, the coverage for many regionally important species falls disproportionately short of the coverage afforded to, say, cereal grains.

Species with Recalcitrant Seeds. Many important crops are poorly represented in *ex situ* collections because their seeds are hard to store or because the species normally propagates vegetatively. Crops such as rubber, cacao, palms, many tuber crops, and many tropical fruits and other tree species can be conserved only *in situ* or in *ex situ* field gene banks. Of the 6,500 potato accessions maintained at the International Potato Center (one of the CGIAR international agricultural research centers) in Peru, 5,000 are clones that are planted each year (Plucknett et al. 1987). Because this procedure is so expensive, the potato is the only root crop with more than 50 percent of its diversity in *ex situ* storage. The development of tissue culture techniques will improve the coverage of species with recalcitrant seeds, but the expense and difficulty of the technique suggests that these species will remain underrepresented in *ex situ* collections.

Wild Species. The principal role of wild crop relatives has been as a source of genes conferring resistance to parasites and pests. More

Table 13. Status of Crop Germplasm Collections

Crop	Number			Coverage (Percent)[a]		Percent Uncollected Types Endangered[b]
	Accessions	Distinct Samples	Collections of 200+ Accessions	Land-races	Wild Species	
Cereals						
Wheat	410,000	125,000	37	95	60	High
Barley	280,000	55,000	51	85	20	Low
Rice	215,000	90,000	29	75	10	2–3
Maize	100,000	50,000	34	95	15	Less than 1
Sorghum	95,000	30,000	28	80	10	100
Oats	37,000	15,000	22	90	50	
Pearl millet	31,500	15,500	10	80	10	High
Finger millet	9,000	3,000	8	60	10	High
Other millets	16,500	5,000	8	45	2	
Rye	18,000	8,000	17	80	30	
Pulses						
Phaseolus	105,500	40,000	22	50	10	Mod-High
Soybean	100,000	18,000	28	60	30	Moderate
Groundnut	34,000	11,000	7	70	50	Mod-High
Chickpea	25,000	13,500	15	80	10	High
Pigeonpea	22,000	11,000	10	85	10	Moderate
Pea	20,500	6,500	11	70	10	
Cowpea	20,000	12,000	12	75	1	High
Mungbean	16,000	7,500	10	60	5	High
Lentil	13,500	5,500	11	70	10	Moderate
Faba bean	10,000	5,000	10	75	15	Moderate
Lupin	3,500	2,000	8	50	5	Mod-High
Root Crops						
Potato	42,000	30,000	28	95	40	90
Cassava	14,000	6,000	14	35	5	Moderate
Yams	10,000	5,000	12	40	5	Moderate
Sweet Potato	8,000	5,000	27	50	1	Moderate
Vegetables						
Tomato	32,000	10,000	28	90	70	
Cucurbits	30,000	15,000	23	50	30	
Cruciferae	30,000	15,000	32	60	25	
Capsicum	23,000	10,000	20	80	40	
Allium	10,500	5,000	14	70	20	
Amaranths	5,000	3,000	8	95	10	
Okra	3,600	2,000	4	60	10	
Eggplant	3,500	2,000	10	50	30	

Table 13. (Cont.)

Crop	Number Accessions	Number Distinct Samples	Number Collections of 200+ Accessions	Coverage (Percent)[a] Land-races	Coverage (Percent)[a] Wild Species	Percent Uncollected Types Endangered[b]
Industrial Crops						
Cotton	30,000	8,000	12	75	20	
Sugar Cane	23,000	8,000	12	70	5	
Cacao	5,000	1,500	12	*	*	
Beet	5,000	3,000	8	50	10	
Forages						
Legumes	130,000	n.a.	47	n.a.	n.a.	
Grasses	85,000	n.a.	44	n.a.	n.a.	

a. Coverage percents are estimates derived from scientific consensus. Coverage of wild gene pool relates primarily to those species in the primary gene pool—those species that were either progenitors of crops, have co-evolved with cultivated species by continuously exchanging genes, or are otherwise closely related.

b. Data unavailable for vegetables, industrial crops, and forages.

* Coverage difficult to estimate because many selections are from the wild.

n.a. Data not available.

Sources: Lyman 1984, Plucknett et al. 1987.

than 20 of the 154 wild species of potato have contributed genes to domestic potatoes (Prescott-Allen and Prescott-Allen 1984, Plucknett et al. 1987). Similarly, wild relatives of tomatoes have been used to breed resistance to at least twelve diseases and one pest that plague the crop (Prescott-Allen and Prescott-Allen 1983). Important contributions of wild germplasm have been made in various other crops, including wheat, rice, barley, cassava, sweet potato, sunflower, grapes, tobacco, cotton, cacao, and sugarcane (Prescott-Allen and Prescott-Allen 1983). The use of wild germplasm is expected to increase as advances are made in biotechnology.

With only two exceptions—wheat and tomatoes—the wild relatives of crops are poorly represented in *ex situ* collections and in very few instances has their *in situ* conservation been attempted. *(See Table 13.)* Thus, many wild relatives of crops of economic importance face the same threat of extinction as other wild species do. The range of the closest wild relatives of maize, *Zea mays* subsp. *mexicana* and *Z. mays* subsp. *parviglumis*, for example, has been halved since 1900 (Wilkes 1972). Similarly, overgrazing is threatening wild relatives of oats, land conversion to agriculture is diminishing the wild germplasm of cassava, and urban development in California has endangered wild relatives of the sunflower (Prescott-Allen and Prescott-Allen 1983).

Livestock. Controlled breeding and the development of livestock varieties suitable for modern commercial production has eroded the genetic diversity in livestock. Throughout Latin America, for instance, the once-dominant Criollo breed of cattle is being replaced by other varieties—in Brazil primarily by Zebu cattle; in Argentina by Angus and Herefords. Similarly, some 241 of the 700 unique strains of cattle, sheep, pigs, and horses that have been identified in Europe are considered endangered (OTA 1987). Response to the threat of genetic erosion of livestock has been slow, partly because genetic improvement has been of less importance in livestock production than in crop production, but also because the expense is comparatively greater.

There is as yet no coordinated effort comparable to the IBPGR for conserving the genetic resources of livestock. Because far fewer species and varieties are involved, less effort than that needed to conserve crop genetic resources is demanded, and though the cost per species will be higher than in plants, the long-term benefits that these genetic resources could provide will be substantial.

VI. Biodiversity Conservation: What are the Right Tools for the Job?

The tools that can be used to maintain biodiversity—the protection of natural or semi-natural ecosystems, the restoration and rehabilitation of degraded lands, and such *ex situ* techniques as zoos, botanic gardens, aquaria, and seedbanks—all provide enormous benefits to humanity. These tools are essential elements of any response to the biotic impoverishment of the planet. Each has its place in a comprehensive strategy for maintaining biodiversity, and their use must be integrated within a larger scheme of land-use planning and management to ensure that the maximum number of species and genes are maintained while meeting peoples' immediate needs.

Land-use Zoning and Protected Areas

Natural and semi-natural ecosystems are the primary reservoirs of the world's biodiversity. In a natural ecosystem, the impact of humanity is no greater than the impact of any other single biotic factor. In semi-natural ecosystems, moderate levels of human disturbance or modification occur. *(See Figure 10.)* How much a specific site can contribute to biodiversity maintenance and which services humanity obtains from the site depends on just how "natural" an ecosystem is. In comparatively unperturbed ecosystems, the characteristic diversity, native species, and existing ecological processes and services can be preserved if the site is large enough and well-managed. Where human dominance is pronounced, certain services (timber production, for instance) may be enhanced while others (species conservation, say) are diminished.

If land is properly managed and zoned, humanity can utilize biological resources without diminishing the biota's capacity to meet future generations' needs. Certain areas can be managed in a natural or semi-natural state with the primary objective of maintaining species diversity and natural ecological processes; others can be managed in semi-natural states primarily for timber production, shifting cultivation, or other services. Land or ocean zoning has already been codified by IUCN and many nations in a system of "protected areas"—legally established areas, under either public or private ownership, where the habitat is managed to maintain a natural or semi-natural state.

Protected areas are among the most valuable management tools for preserving genes, species, and habitats and for maintaining various ecological processes of importance to humanity. Such protected areas may or may not include human settlements, transportation systems, and subsistence or commercial use. All protected areas require some intervention in the ecosystem—whether boundary patrol or the supervision of recreation, or resource extraction. Various man-made components may be added: research stations, monitoring devices, roads, or logging equipment.

Figure 10: Ecosystems Subject to Various Degrees of Human Modification

Natural   Semi-Natural   Urban

A natural ecosystem is one in which the impact of humanity is no greater than that of any other single biotic factor. Semi-natural ecosystems are subject to moderate levels of human disturbance or modification and encompass a broad range of conditions between urban environments or intensive agriculture and natural environments.

Protected areas have one of two major management objectives. *(See Appendix 3—Protected Area Objectives.)* Protected areas established primarily to maintain biological diversity and natural formations are referred to as ''strictly protected areas'' (IUCN categories I thru III): scientific reserves, national parks, and natural monuments (Miller 1975, IUCN 1984). In contrast, other categories (IUCN categories IV thru VIII) focus on controlled resource exploitation while retaining limited but significant commitments to maintaining biodiversity: managed nature reserves, protected landscapes, resource reserves, anthropological reserves, and multiple-use areas (including game ranches, recreation areas, and extractive reserves).

Although protected areas can provide selected goods and services on a long-term basis, various management objectives are not always compatible. Where the preservation of

species and habitats is not a primary management objective, some loss of species or genetic diversity may occur. Nevertheless, relative to the surrounding urban or intensively altered landscape, *any* category of protected area helps maintain biodiversity and ecological processes. Particularly in protected areas where conserving species and habitats is a secondary objective, the challenge is to make sure that management strategies developed to achieve primary objectives are as compatible as possible with the need to maintain the various species and habitats.

Protected areas must be conceived and managed as a *system* of protected sites—no one of which meets all of the possible objectives of protected area establishment, but which together provide the essential services that humanity requires, including the maintenance of the world's biological diversity. Management within the context of the entire system of protected areas also helps ensure the viability of the species populations that they contain. In particular, this approach makes it easier to identify the need for conservation corridors linking protected areas, and it allows more complete assessments of species' survival prospects.

Protected areas must be conceived and managed as a **system** *of protected sites—no one of which meets all of the possible objectives of protected area establishment but which together provide the essential services that humanity requires from natural and semi-natural ecosystems.*

Status of Protected Areas

Some 4,500 areas covering 485 million hectares—3.2 percent of the earth's land surface—have been designated as legally protected scientific reserves, national parks, natural monuments, managed nature reserves, and protected landscapes. *(See Table 14.)* Of these areas, 307 million hectares, 2 percent of the land surface, are strictly protected. Excluding Greenland, which contains the world's largest national park (700,000 km^2), only 1.6 percent of the world's land surface is strictly protected. In some countries, more land is strictly protected than the total of IUCN categories I thru III would suggest because certain areas managed for extractive or multiple-use purposes *contain* strictly protected wildlands designated through zoning, such as wilderness areas and research areas within some U.S. national forests.

Strictly protected areas will never cover more than 5 to 10 percent of the earth's land surface. National forests and other protected areas

where resource extraction takes precedence over maintaining biodiversity probably won't cover more than an additional 10 to 20 percent of the earth's surface. In the United States, national forests account for 8.5 percent of the land surface, and another 15 percent is managed by the Bureau of Land Management for various uses, including the maintenance of natural or semi-natural ecosystems. In Costa Rica, some 17 percent of the land is managed as forest and Indian reserves.

During the past 20 years, more protected areas have been established than ever before. *(See Table 15.)* Indeed, more area has been given protected status since 1970 than in all previous decades combined. Protected areas in IUCN categories I to V now exist in 124 countries and in all of the world's biogeographic realms. *(See Box 5.)* However, biogeographic coverage is incomplete. Fifteen biogeographic provinces have no protected areas, and 30 have five or fewer protected areas that together encompass less than 1,000 square kilometers.

Detailed reviews of the coverage of protected areas, the extent of habitat loss, and priorities for establishing protected areas have been made by IUCN for the Afrotropical, Indo-Malayan, and Oceanian biogeographic realms (IUCN/UNEP 1986a, b, c). In addition, the U.S. National Academy of Sciences reviewed the threatened tropical forest sites of outstanding biodiversity value in 1980 (NRC 1980), and a recent survey of ten critical tropical forest sites—or "hotspots"—has been published by Myers (1988). *(See Table 16.)* Some differences exist in the areas identified in these reviews, in part due to differences in the geographic units being analyzed, but their authors in general agree on which regions should receive highest priority. Detailed national reviews that consider the adequacy of existing areas, the needs of local people, and the coverage of protected areas are needed, but these general reviews are useful in identifying global priorities.

Just as important as the size of the area protected is the extent to which funding has

Table 14. Coverage of Various Biogeographic Realms by Protected Areas[a]

| | Protected Area Designation (IUCN Category Number) | | | | | | | | | | | |
| | Scientific Reserve (1) | | National Park (2) | | Natural Monument (3) | | Wildlife Reserve (4) | | Protected Landscape (5) | | Total | |
Realm[b]	No.	km²	No.	km²	No.	km²	No.	km²	No.	km²	No.	km²
Nearctic	5	11,600	142	1,155,500[c]	32	64,200	259	380,700	40	113,600	478	1,725,600
Palearctic	313	273,100	204	112,300	24	2,000	649	172,800	494	171,700	1684	731,900
Afrotropical	23	17,600	152	574,300	1	<100	260	268,700	8	300	444	860,900
Indomalayan	63	27,900	158	111,300	5	300	411	180,400	39	2,900	676	322,800
Oceanian	17	25,900	10	3,300	0	0	24	19,600	1	<100	52	48,900
Australian	58	23,100	248	192,600	0	0	277	137,500	40	3,700	623	356,900
Antarctic	29	6,500	11	21,000	5	200	85	3,500	0	0	130	31,200
Neotropical	55	63,900	224	423,900	22	2,900	126	244,900	31	32,500	458	768,100
Total	563	449,600	1,149	2,549,200	89	69,600	2,091	1,408,100	653	324,800	4,545	4,846,300

a. Sites included are all those protected areas over 1000 hectares, classified in IUCN Management Categories 1 thru 5, that are managed by the highest accountable authority in the country.

b. Realms follow the definition of Udvardy 1975.

c. The Greenland National Park of 700,000 square kilometers has a significant effect on comparative statistics. It is an order of magnitude larger than any other single site.

Source: Protected Areas Data Unit, World Conservation Monitoring Centre, May 1989.

allowed existing protected areas to achieve their objectives. Where threats to a site are minor, designating an area as protected without managing it intensively may be enough to maintain the site's biodiversity. But where threats are greater and intensive management is necessary, protected areas commonly lack the necessary political and financial support (Machlis and Tichnell 1985). In 84 areas listed by IUCN's Commission on National Parks and Protected Areas as "threatened" (23 in Africa, 18 in Asia, 3 in Australia and the Pacific, 17 in Europe, 20 in Latin America, and 6 in North America), poaching, mining, settlement, military activities, acid rain, and other wide-ranging threats have thwarted stated conservation objectives, and many more than just these 84 areas need increased support. In one survey of 98 national parks, fully 73 percent of the parks reported that they were understaffed (Machlis and Tichnell 1985), and many protected areas have trained staff who don't have the funds needed for fuel, equipment, or subsistence.

Potential Conservation Importance

How much of the world's species and genetic diversity exists today on the small fraction of the earth's surface that we can expect to remain in a natural or semi-natural state? The following examples give some idea:

- Approximately 88 percent of Thailand's resident forest bird species occur in its national parks and wildlife sanctuaries which currently cover 7.8 percent of its land area (Round, P.D. 1985, cited in IUCN/UNEP 1986a).

70

Table 15. Global Rate of Protected Area Establishment

Decade	Number of Areas	Size (km²)
Unknown Date	711	194,395
Pre-1900	37	51,455
1900–1909	52	131,385
1910–1919	68	76,983
1920–1929	92	172,474
1930–1939	251	275,381
1940–1949	119	97,107
1950–1959	319	229,025
1960–1969	573	537,924
1970–1979	1,317	2,029,302
1980–1989	781	1,068,572

Notes: Includes all areas over 1,000 hectares in IUCN management categories 1-5 protected by the highest accountable authority in each nation (i.e., state parks and reserves are not included).

• Populations of all of the bird and primate species in Indonesia will be contained within the existing and proposed reserve systems. (Some 3.5 percent of Indonesia is protected; a further 7.1 percent is proposed.) (IUCN/UNEP 1986a).

• In 11 African countries with reasonably complete accounts of their avifauna, more than three quarters of bird species occur in existing protected areas. *(See Table 17.)* Sayer and Stuart (1988) estimate that for Africa as a whole more than 90 percent of tropical forest vertebrates would be maintained if a few critical sites were added to the existing protected areas and adequate management and financial support provided.

• In a survey of 30 protected areas in Africa, only 3 of the 70 African species in the Coraciaformes family of birds (kingfishers, bee-eaters, rollers, hoopoes, hornbills) are not found in any areas, 90 percent are found in more than one protected area, and 55 percent in more than five (IUCN/UNEP 1986b).

Box 5. Biogeographic Classifications

Any region's species, habitats, and ecosystems are influenced by the region's history and the characteristics of the physical environment—particularly the soils, landforms, and climate—and the world can be divided into a number of regions with broadly similar flora and fauna. Biogeographers working on terrestrial ecosystems, for example, generally identify eight biogeographic *realms* in which the species within most taxa are more closely related to each other than they are to species in other regions (Pielou 1979). *(See Figure 11.)* The boundaries between these realms (frequently, oceans or deserts) generally block the dispersal of organisms. These biogeographic realms can be subdivided into smaller units, known as biogeographic *provinces*, that reflect environmentally determined differences in how individual species or groups of species are distributed.

The most widely used such system was developed in the 1970s by Udvardy (1975) for the terrestrial environment based on initial work of Dasmann (1973). Udvardy subdivided the eight biogeographic realms into 193 provinces (later increased to 227; Udvardy 1984). Used by the Man and the Biosphere Program of UNESCO and in IUCN's global inventory of protected areas, the Udvardy system describes general biogeographic patterns, but the system is not fine-tuned enough to meet most national and local resource-planning needs.

Box 5. (Cont.)

A more recent world-wide biogeographical mapping approach, one that generally follows the approach of Dasmann and Udvardy, is based on ''ecoregions.'' These regions are defined on the basis of both climate and vegetation indicators and divided into ''sections'' within provinces that differ vis-à-vis the dominant vegetation (Bailey and Hogg 1986). In the United States, the ecoregion classification scheme is being used by the Fish and Wildlife Service, the Forest Service, and other land-management agencies to monitor acid rain effects and wetlands loss.

The ecoregion approach roughly doubles the resolution of the Udvardy classification scheme. For example, 31 ecoregion provinces are identified for the United States as compared with 16 biogeographical provinces. A more fine-grained classification still is that used widely in Africa: the UNESCO vegetation map of Africa (White 1983). The Holdridge Life Zone System, which is also quite detailed, is often used in the Neotropical

Realm. Still more fine-grained schemes are possible at national levels. For example, a recent vegetation map of Venezuela identifies 150 vegetation types (Huber and Alarcon 1988).

Duplicating the terrestrial scheme of biogeographic classification in coastal and marine environments exactly is impossible. Barriers to dispersal have not been as significant in determining the distribution of organisms in the marine environment as on land, and the three-dimensionality of the ocean is more striking and harder to map than that of most terrestrial environments. These difficulties notwithstanding, Briggs (1974) identified 23 marine biogeographical regions based on the distribution of groups of marine animals with evolutionary affinities. Hayden et al. (1984) combined this zoo geographical approach with considerations of the physical properties of marine systems (currents, climate, and physical location—coastal, shelf, open ocean, island) to identify 40 marine biogeographic provinces. *(See Figure 12.)*

• In Costa Rica, 55 percent of the sphingid moths that occur in the country have breeding populations in Santa Rosa Park (0.2 percent of the land area), and when Santa Rosa is enlarged and becomes the 82,500-hectare Guanacaste Park (1.6 percent of the country), nearly all the 135 species will be covered (Janzen et al., unpublished manuscript).

The extent to which protected areas ensure the survival of species in these examples is overstated to the extent that some species occurring in the protected areas are present in numbers below their minimum viable population size. Also, plant species present a special case: since plants tend to exhibit high habitat specificity and local endemism, comparatively

more habitats must be represented in protected areas to adequately maintain plant diversity. For example, of all plant species listed as endangered and vulnerable in Australia, only 21 percent occur in national parks or proclaimed reserves (Leigh et al. 1982). In contrast, these examples understate the overall status of a nation's species to the extent that some species outside of protected areas may have viable populations in disturbed landscapes. For example, 39 percent of the 70 bird species not recorded in parks and sanctuaries in Thailand are open country birds that can survive further deforestation (Round, P.D. 1985, cited in IUCN/UNEP 1986a).

These examples do not suggest that the coverage afforded by existing protected areas is

Table 16. Priority Regions for Conservation of Biological Diversity in Neotropics, Afrotropics, Indo-Malaysia, and Oceania

Region	National Academy of Sciences (NRC 1980)[a]	IUCN/UNEP (1986a,b,c)[b]	Myers (1988)[c]
Latin America		(not reviewed)	
Coastal forests of Ecuador	x		x
Atlantic coast, Brazil	x[d]		x
Eastern and southern Brazilian Amazon	x		
Uplands of western Amazon			x
Colombia Choco			x
Africa			
Cameroon	x		
Mountainous regions of East Africa	x[e]	x	
Madagascar	x	x	x
Equatorial West Africa		x	
Sudanian zone[f]			
Asia			
Sri Lanka	x	x[g]	
Indonesian Borneo (Kalimantan)	x		
Northern Borneo (Sarawak)			x
Sulawesi	x		
Eastern Himalayas		x	x
Peninsular Malaysia			x
Philippines		x	x
Indo-China[h]		x	
Bangladesh/Bhutan/Eastern Nepal		x	
Vietnam/Coastal Kampuchea and Thailand		x	
Southeast China		x	
Oceania			
New Caledonia	x		x
Hawaii	x		

a. Priority based on moist tropical forest sites with: great biological diversity, high local endemism, and a high conversion rate to other purposes.

b. Priority based on tropical forests sites with: exceptional concentrations of species, exceptional levels of endemism, and exceptional degrees of threat.

c. Priority based on biogeographic units with: small percentage of land included in protected areas; enough natural habitat remaining that the protected area system could be expanded; high richness of plant, mammal, and bird species; and high levels of endemism.

d. Southeast extension of the state of Bahia between the Atlantic coast and 41°31'W longitude and between 13°00'W and 18°15'S latitude; as well as a small area near Linhares, farther south in the state of Espíritu Santo.

e. Moist tropical forests of the Usambara, Nguru, and Uluguro hills of Tanzania and associated ranges and the related montane forests in Kenya.

f. Band immediately south of the Sahel, from about 5° to 15°N latitude, stretching east to Ethiopia.

g. Southwestern Sri Lanka.

h. This biogeographic unit includes parts of interior Burma, Thailand, and Kampuchea, most of Laos and part of southern China.

Figure 11: Terrestrial Biogeographic Provinces

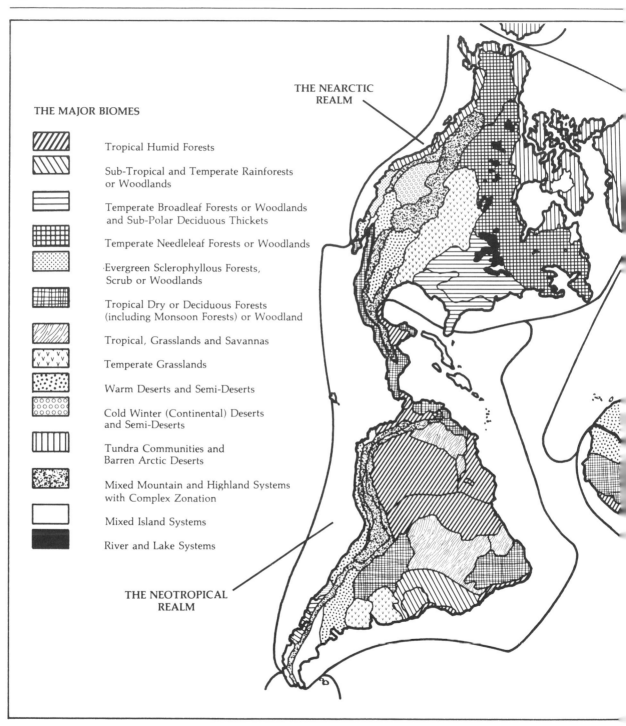

THE MAJOR BIOMES

	Tropical Humid Forests
	Sub-Tropical and Temperate Rainforests or Woodlands
	Temperate Broadleaf Forests or Woodlands and Sub-Polar Deciduous Thickets
	Temperate Needleleaf Forests or Woodlands
	Evergreen Sclerophyllous Forests, Scrub or Woodlands
	Tropical Dry or Deciduous Forests (including Monsoon Forests) or Woodland
	Tropical, Grasslands and Savannas
	Temperate Grasslands
	Warm Deserts and Semi-Deserts
	Cold Winter (Continental) Deserts and Semi-Deserts
	Tundra Communities and Barren Arctic Deserts
	Mixed Mountain and Highland Systems with Complex Zonation
	Mixed Island Systems
	River and Lake Systems

THE NEARCTIC REALM

THE NEOTROPICAL REALM

Areas demarcated by lines within continents indicate separate biogeographic provinces. Biogeographic provinces are subdivisions of biological realms.

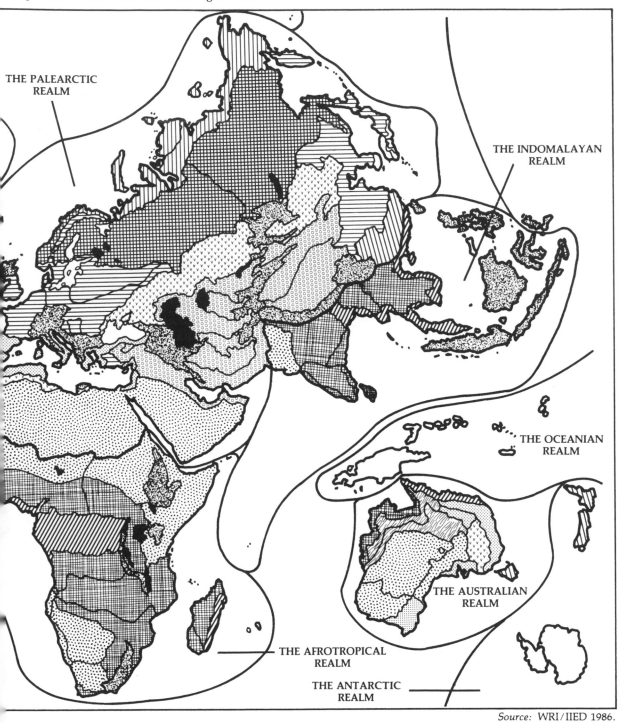

THE PALEARCTIC
REALM

THE INDOMALAYAN
REALM

THE OCEANIAN
REALM

THE AUSTRALIAN
REALM

THE AFROTROPICAL
REALM

THE ANTARCTIC
REALM

Source: WRI/IIED 1986.

Figure 12: Marine Biogeographic Provinces

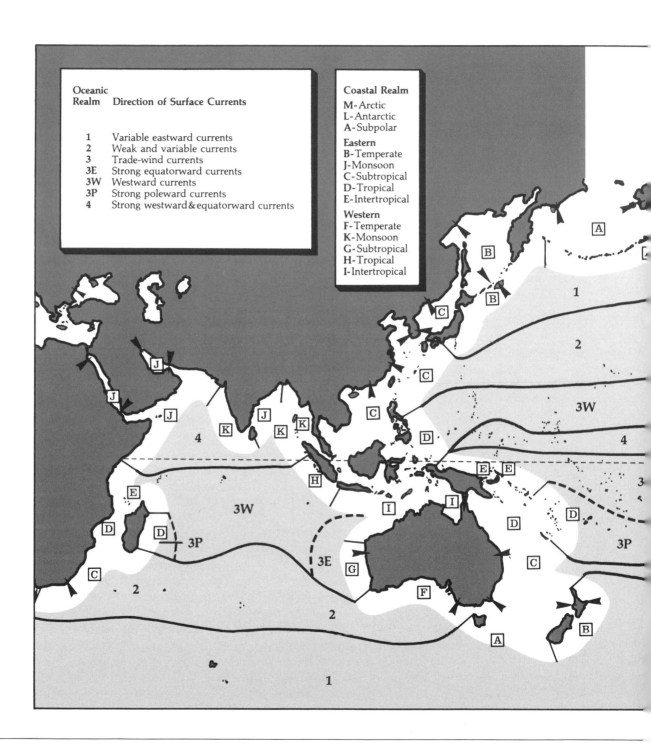

Oceanic realms are shaded, Coastal Realms are clear, and Faunal Provinces are bounded by arrows along coast. Faunal Provinces are not always subdivisions of Coastal Realms. Not shown are Marginal Seas and Archipelagos: shallow seas (e.g., Caribbean, Mediterranean) situated between the coastal margins and the continents or between two coastal margins and named for the Coastal Realm in which they are contained.

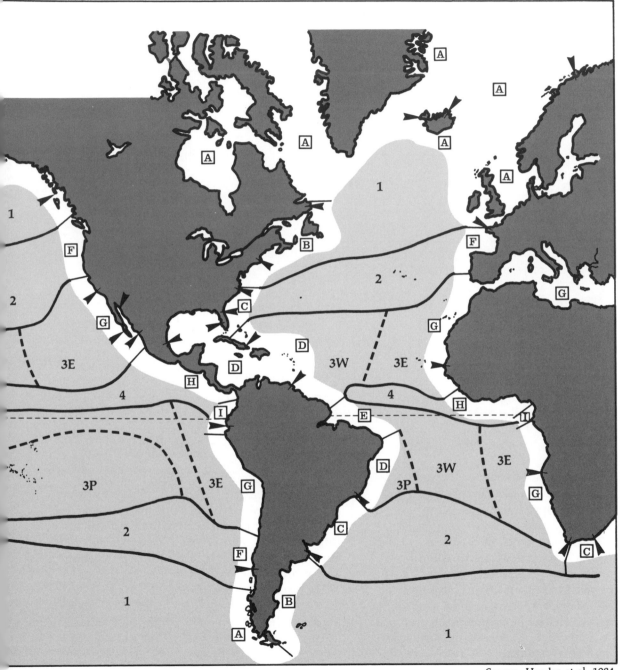

Source: Hayden et al. 1984.

Table 17. Species Occurring in Protected Areas in Selected African Countries

Country	Percent of National Land Area Protected (WRI/IIED 1988)[a]	Number of Bird Species in Country	Percent in Protected Areas
Cameroon	3.6	848	76.5
Ghana	5.1	721	77.4
Côte d'Ivoire	6.2	683	83.2
Kenya	5.4	1064	85.3
Malawi	11.3	624	77.7
Nigeria	1.1	831	86.5
Somalia	0.5	639	47.3
Tanzania	12.0	1016	82.0
Uganda	6.7	989	89.0
Zaire	3.9	1086	89.0
Zambia	8.6	728	87.5
Zimbabwe	7.1	635	91.5

a. Includes IUCN categories 1 to 5 (scientific reserves, national parks, natural monuments, managed nature reserve, protected landscapes).

Source: Sayer and Stuart 1988.

sufficient, but they do indicate that the relatively small land area allocated to protected areas can be extremely important if the areas are well-chosen and well-managed. If either of these conditions is not met, protected areas will not be nearly as useful as they could be.

Strictly protected areas are not the only sites with considerable potential for conserving biodiversity. With proper incentives, lands managed for extractive uses can contribute greatly to the maintenance of biodiversity. Well managed semi-natural ecosystems could both provide habitat for diverse local species and provide products and services to local people (Johns 1985). For example, much of Amazonia's wildlife can survive in habitat that is slightly or moderately disturbed by, say, low-density shifting agriculture and low levels of selective logging (Johns 1986, 1988). Similarly, in Malaysia, a 25-year-old logged forest

contained nearly 75 percent of the avifauna of a virgin forest (though the presence of adjacent undisturbed forest probably increased the number found in the disturbed site) (Wong 1985). Thus, forest management for timber extraction need not create a biological wasteland.

Numerous factors, including financial support, management capacity, and political stability influence whether or not a protected area can realize its potential to maintain biodiversity. But the potential itself is largely determined at the outset by the choice of location, the size of the area, and by the zoning and management of surrounding lands or waters.

Reserve Size, Network Design, and Transition Zones

In larger protected areas, a higher proportion of the species found at a given site will be maintained. But while habitat size is extremely important, so is the choice of which habitats to protect. Establishing more small protected areas in a variety of habitats may save more species than establishing fewer larger protected areas would, since the smaller areas will provide a larger sample of different assemblages of organisms.

Several examples that demonstrate the need to consider both the size of a protected area and the number of habitats covered in the design of a protected area network are found in North America. The total number of large mammals present in three protected areas that represent three different habitats (Redwood, North Cascades, and Big Bend national parks) exceeds the number in the single largest North American park (Yellowstone National Park), even though the total area of the three is only 57 percent that of Yellowstone (Quinn and Harrison 1988). Similarly, a greater diversity of prairie plant species can be conserved in a group of small sites in the mid-western United States than can be saved in a single large site of the same total area (Simberloff and Gotelli 1984). Conserving populations in several small protected areas rather than a single large area also reduces the risk of extinction caused by a local environmental catastrophe, such as disease, fire, or severe weather (Simberloff and Abele 1982, Quinn and Hastings 1987).

Considering that both land and funds for establishing and managing protected areas are limited, balancing the detrimental effect of decreased size of protected areas with the beneficial effect of increased habitat coverage afforded by a larger number of smaller areas is critical. This question of balance led to a heated scientific debate in the late 1970s and early 1980s. Eventually, scientists agreed that a combination of large and small protected areas would best meet the twin objectives of conserving biological diversity and providing services important to humanity (Soulé and Simberloff 1986).

Protected areas need not become islands in a sea of disturbed land. Corridors of natural or semi-natural habitat can link protected areas together and thereby increase the breeding population size in one reserve by joining it to a population in an adjacent reserve (Simberloff and Cox 1987). Those with a north-south orientation or located along elevational gradients facilitate seasonal migration and may allow species distributions to shift in response to global warming. La Zona Protectora, a 7,700-hectare stretch of forest 3 to 6 km wide linking Braulio Carillo National Park with La Selva Biological Station in Costa Rica, preserves a migration corridor for approximately 35 species of birds plus an unknown number of mammals and insects (Wilcove and May 1986).

Protected areas need not become islands in a sea of disturbed land.

Conservation corridors have some potential drawbacks. First, populations linked by corridors may be more susceptible to extermination by disease or fires (Simberloff and Cox 1987). Second, the economic cost of establishing and maintaining corridors can be high: developed land sometimes has to be purchased and the high ratio of edge to area in the corridor increases the management cost per unit area. Consequently, scarce conservation funding may often be more efficiently spent for other purposes, such as managing existing protected areas better.

Within a context of regional land management, the interface between areas with differing management objectives—for example, between a national park and agricultural land, or between a national park and a national forest—assumes considerable importance. A well managed transition zone can safeguard the lands surrounding a protected area from such effects as damage caused by large animals feeding on the neighboring gardens and crops.

Proper management can also minimize potentially degrading disturbances, such as pesticide run-off from surrounding lands. Yet, transition zones are often neglected, usually because they don't fall under the protected area manager's purview nor that of the neighboring landholder. This neglect diminishes the protected area's effectiveness and can create conflicts between adjacent land uses and protected area management.

Historically, a transition zone was referred to as a "buffer zone" and considered a collar of land managed to filter out inappropriate influences from surrounding activities. Typically, neighboring lands—such as national forests—were expected to be managed to meet the needs of the strictly protected area, but with little incentive for such management it has rarely happened that way. Today, the challenge is to incorporate the management of transition zones into protected area planning and financing. For example, the Indonesian park department has established tree plantations around segments of Baluran and Ujung Kulon national parks in Java to meet neighboring people's fuel wood needs while minimizing disturbance to the protected area.

Ecological Restoration and Rehabilitation

The rehabilitation of degraded lands and the restoration of ecosystems have become increasingly important elements of resource conservation throughout the world. *Rehabilitation,* which aims to revive important ecological services on degraded lands, is becoming particularly important in mountainous regions, arid lands, and irrigated crop lands. In mountainous areas, the loss of forest and other vegetative cover has often increased soil erosion. Arid lands have suffered declining soil fertility and increased erosion as a result of agriculture and overgrazing. Resource productivity has declined on an estimated 80 percent of rangelands and 60 percent of croplands in arid regions of developing countries, and irrigated croplands have been degraded by salinization,

waterlogging, and alkalinization (WRI/IIED 1988).

More ambitiously, *restoration* attempts to bring lands modified by human use back to their natural state. Because determining the "pre-disturbance" state of most ecosystems is difficult and because ecosystems continually change, complete restoration is rarely a realistic goal. But the approximate re-creation of natural communities is becoming central to efforts to maintain biodiversity and restore important ecological services.

Both restoration and rehabilitation make use of physical and biological interventions. Physical means include drainage systems in waterlogged lands and check dams or contour plowing to slow erosion rates, while biological interventions include growing grasses to slow erosion, nitrogen-fixing trees to increase the nutrient content of soils, drought-adapted trees for hillside reforestation, and so forth. Rehabilitation often makes use of exotic species, since the primary goal is to restore critical ecological services rather than the natural community. In contrast, restoration attempts to restore the natural complement of species.

Ecosystem restoration does not always require interventions, however. Left to natural processes, many ecosystems will return to something like their pre-disturbance condition if populations of the original species still exist nearby. How long natural recovery takes depends upon the type of ecosystem and the type of disturbance. In Brazil's caatinga forest, natural recovery of slash-and-burn agricultural sites requires more than a century, but sites cleared by bulldozer may take 1,000 or more years to recover (Uhl et al. 1982). Similarly, sites in tropical lowland wet and dry forest require an estimated 1,000 and 150 years, respectively, to recover from timber harvest (Opler et al. 1977).

If an ecosystem has been physically transformed or if pre-disturbance species cannot disperse to the site, natural processes alone won't restore it. Numerous lakes throughout the world and the prairies of central North America have been fundamentally changed by introduced species and can't return to their natural states unless the exotic species are removed. Elsewhere, soil erosion, salinization, or the loss of mycorrhizal mutualists has changed ecological systems so radically that native species can't become established without such interventions as seeding, planting, inoculation of soils with mycorrhizal fungi, and weed, fire, or predator control (R. Miller 1987).

But even where communities and ecosystems might naturally revive after disturbance, restoration technologies can speed recovery. In recent years, for instance, the recovery (and creation) of numerous coastal salt marshes has been significantly accelerated by planting common salt marsh species (Jordan et al. 1988).

As noted by Jordan et al. (1988), restoration ecology falls in an "economic and psychological no-man's land, appealing to neither agricultural interests nor to environmentalists concerned with the preservation of pristine natural areas." Yet, restoration ecology can enhance *in situ* conservation. Where existing protected areas are too small to maintain certain species, ecological restoration can be used to enlarge areas of natural habitat or to establish conservation corridors between reserves.

The planned 800-km² Guanacaste National Park in Costa Rica will be created largely through ecological restoration. The park is located in a region formerly covered by tropical dry forest—the most threatened of all major lowland tropical forest habitats (Janzen 1986, 1988b). Some 700 square kilometers is now being restored from pasture and forest fragments. This restoration will help save many of the dry forest species from extinction, boost the local economy, and help re-establish the cultural ties of the people to their land. The restoration will rely primarily on natural processes because remaining remnants of forest can provide seeds to the restored sites; however, to speed the process, fires will be suppressed and some

cattle may be allowed to graze initially since reducing grass height enhances tree seedling survival (Janzen 1986, 1988b).

In Guanacaste Park, the remaining fragments of natural communities are big enough that the region's native species are still present. But in many restoration programs, *in situ* and *ex situ* conservation efforts must be closely coordinated. Populations of endangered species are increasingly being maintained in captivity only until suitable wild habitat has been restored. As part of the recovery program for the red wolf in the United States, an effort was made in the mid-1970s to capture all of the remaining wild wolves. Captive breeding populations were then established while scientists looked for suitable habitat for the wolves in their natural range. Since 1986, the wolves have gradually been returned to one of these habitats (Phillips and Parker 1988).

Just as restoration helps conserve biodiversity, rehabilitation and restoration programs require the very species that biodiversity conservation seeks to maintain. For instance, rehabilitation projects are benefiting from the more widespread use of species that can grow under stressful environmental conditions, and both restoration and rehabilitation draw upon species maintained in protected areas, zoos, botanic gardens, and seedbanks to establish populations.

This current focus on stressful environmental conditions is not misplaced. As the threat of climate change to biodiversity grows, ecological restoration and rehabilitation will become increasingly important. Along with habitat fragmentation, rapid climate change will prevent many species from migrating to more suitable climates. Increasingly, restoration techniques will be used to introduce species to new ranges and to accelerate the re-establishment of such communities.

Ex Situ Conservation

While the importance of *in situ* conservation cannot be overemphasized, *ex situ* conservation in zoos, aquaria, botanic gardens, and germplasm banks both complements *in situ* techniques and serves many other purposes—among them, maintaining viable populations of species threatened in the wild, providing educational and public awareness services, and serving as sites for basic and applied research. Moreover, in some regions the threats to species survival are so severe that no hope exists for their long-term *in situ* maintenance.

Zoos and Aquaria

There are roughly 500,000 mammals, birds, reptiles, and amphibians in captivity in zoos throughout the world (Conway 1988). Zoos contribute in many ways to the conservation of biodiversity. They propagate and reintroduce endangered species, they serve as centers for research to improve management of captive and wild populations, and they raise public awareness of biotic impoverishment (Foose 1983).

The contributions that zoos have already made to the conservation of biodiversity are dramatic. Zoo populations are now the only representatives of several species, including the California condor *(Gymnogyps californianus)* and possibly the black-footed ferret *(Mustela nigripes)*, and at least 18 species have been reintroduced into the wild after captive propagation (Conway 1988). In at least six cases—Pere David deer *(Elaphurus davidianus)*, Przewalski horse *(Equus przewalski)*, red wolf *(Canis rufus)*, Arabian oryx *(Oryx leucorx)*, American Bison *(Bison bison)*, Guam kingfisher *(Halcyon cinnamomina cinnamomina)*, and Guam rail *(Rallus owstoni)*—the species were extinct in the wild at the time of reintroduction—extinct for some 800 years in the case of the Pere David deer! (U. Seal, Species Survival Commission IUCN, personal communication, February 1989). Numerous other species not completely lost from the wild, such as the peregrine falcon *(Falco peregrinus)* and the alpine ibex *(Capra ibex ibex)*, have been reintroduced to portions of their range from which they were extirpated.

The potential role of zoos as a site for preserving species over the long term is limited by both space and expense, particularly in the case of vertebrates. In the United States, zoos contain self-sustaining populations of a total of only 96 species (some of them threatened) (Ralls and Ballou 1983). Even if roughly one-half of the spaces in all the world's zoos were suitable for captive propagation of threatened species and if 500 members of each species was maintained, only 500 species could be kept alive and healthy in captivity.

This sobering calculation aside, for two reasons the contribution of zoos to biodiversity conservation is likely to considerably exceed this figure. First, zoo managers generally consider a smaller population—roughly 100 to 150 individuals per species—enough for long-term captive propagation (Conway 1980). Deleterious inbreeding effects are still reasonably rare at this population size, and, with proper genetic management, more than half of the genetic diversity of the species can be preserved for at least 100 generations. At these smaller population sizes, zoos could sustain roughly 900 species (Conway 1988). Second and more important, zoos with endangered species are shifting their emphasis from long-term holding to the return of animals to the wild within two or three captive generations (U. Seal, personal communication, March 1989). A number of endangered species, including the black-footed ferret, red wolf, Guam rail, Puerto Rican parrot *(Amazona vittata)*, Mauritius pink pigeon *(Nesoenas mayeri)*, and the whooping crane, are being managed through such programs.

Until recently, the role of aquaria in the captive propagation of threatened species has been less important than that of zoos. However, given the growing threats to freshwater species, the need to enhance the role of aquaria as *ex situ* conservation tools is clear. Accordingly, the Captive Breeding Specialist Group of IUCN is mounting a major effort to develop captive breeding programs for endangered fish species, starting with the fishes of Lake Victoria, the desert fishes of North America, and Appalachian stream fishes (Les Kaufman, New England Aquarium, personal communication,

June 1989). In this program, natural habitat will be restored and—to insure against the loss of wild species and help educate the public on threats to fishes—species will be propagated in aquaria.

Because the world stands to lose a higher proportion of large vertebrates than of other taxa, some disproportionate expenditure on their conservation is justified.

Most captive propagation programs in zoos focus on vertebrates, partly because zoos have historically been display facilities, but also because the extinction threats to vertebrates are well known. This seemingly lopsided focus on a small number of large vertebrates ("charismatic megavertebrates") makes some sense even though, comparatively, far more invertebrates face extinction. The current extinction crisis is almost exclusively terrestrial, and top carnivores and large animals in general will be lost out of proportion to their representation in the biota. *(See Box 4.)* Because the world stands to lose a higher proportion of large vertebrates than of other taxa, some disproportionate expenditure on their conservation is justified. Consider too that large vertebrates often get the public interested in general in protecting habitats—with all of their attendant species. Still, the expense of propagating large vertebrates in captivity can be considerable, and the same investment in *in situ* conservation could maintain a much larger number of smaller species of invertebrates and plants. The question here is whether public financial support for less "glamorous" species would even come close to that obtained for the large vertebrates.

The potential contribution of zoos to conserving species and genetic diversity could be enhanced considerably if more research on

captive propagation and reintroduction techniques were carried out. Right now, zoos' success in maintaining populations of endangered species is limited. Only 26 of 274 species of rare mammals in captivity are maintaining self-sustaining populations (Ralls and Ballou 1983). Fortunately, this state of affairs can improve. Ongoing research has already led to significant advances in technologies for captive propagation, including artificial insemination, embryo transfer, and the genetic management of small captive populations (Dresser 1988). Germplasm storage techniques, including long-term storage of embryos, have also improved considerably. Embryo transfer, in particular, has a tremendous potential for use in captive propagation since it allows the introduction of new bloodlines into captive populations without transporting adults—and any diseases they might carry—between zoos or between wild populations and zoos. Long-term cryogenic storage of embryos has become almost routine for some species—including the mouse, rabbit, and cow—but the technology is only beginning to be experimentally applied to captive species in zoos.

Other new findings and techniques also hold promise. Improved management of the genetics of small captive populations has also made propagation easier by making headway against some of the problems associated with inbreeding, including decreased fertility and high juvenile mortality (Ralls and Ballou 1983). Advances in the theory of population genetics and improved data management are now improving the genetic management of captive populations. In 1973, the computerized International Species Inventory System (ISIS) was developed to catalog genealogical information on individual animals in some 326 zoos (N. Flesness, Minnesota Zoological Garden/ISIS, personal communication, March 1989). Matings can thus be arranged by computer to ensure that genetic diversity is preserved and inbreeding is minimized.

These advances aside, the lack of detailed genetic information on both captive and wild populations bedevils zoo managers. In the early 1980s, for example, the five remaining individuals of the dusky seaside sparrow were crossed with what was thought to be their closest relative—Scott's seaside sparrow—to produce hybrids with genetic compositions ranging from 50 to 87.5 percent dusky. But mitochondrial DNA analysis of the relationship of the dusky seaside sparrow to other subspecies later showed that the Scott's seaside sparrow is a relatively distant relative while the dusky is extremely closely related to four other subspecies (Avise and Nelson 1989). More thorough surveys of the genetic variation of both captive and wild populations will make such slip-ups in breeding programs in zoos and aquaria less common.

Botanic Gardens

It is much easier to maintain captive populations of plants than of animals (Ashton 1988). Plants require less care than animals; their habitat requirements can generally be provided more easily; caging is unnecessary; individuals can be crossed more readily; many can be self-pollinated or vegetatively propagated; and most are bisexual, which means that only half as many members of a species are needed to maintain genetic diversity. Moreover, as mentioned already, many plants can be readily preserved during their dormant (seed) stage. For all these reasons, botanic gardens are extremely important tools for maintaining species and genetic diversity.

The world's roughly 1,500 botanic gardens today contain at least 35,000 plant species or more than 15 percent of the world's flora, with estimates ranging as high as 70,000 to 80,000 species (Raven 1981; IUCN/WWF 1989; P. Raven, personal communication, April 1989). The Royal Botanic Gardens of England (Kew Gardens) alone contains an estimated 25,000 species of plants (10 percent of the world's flora); and IUCN considers some 2,700 of these species rare, threatened, or endangered (G. Prance, Royal Botanic Gardens, personal communication, May 1989).

Even if only 300 to 400 of the world's botanic gardens are able to hold major conservation collections and only 250 of these maintain seedbanks, it would still be possible to save viable populations of up to 20,000 plant species from extinction through the botanic garden system.

For specific taxa, the coverage afforded by botanic gardens is even higher. One collection in California contains 72 of the 110 known species of pines (*Pinus* spp.), one botanic garden in South Africa contains roughly one quarter of the country's flora, and one garden in California boasts one third of the state's native species (Raven 1981). In at least one case, a plant species extinct in the wild (*Clarkia franciscana*) has been preserved in a botanic garden and reintroduced to its native habitat in California (Raven 1976). Even if only 300 to 400 of the world's botanic gardens can harbor major conservation collections and only 250 of these maintain seedbanks, by one estimate it would still be possible to save viable populations of up to 20,000 plant species from extinction in them (IUCN/WWF 1989).

Botanic gardens' potential role in the conservation of biodiversity is exemplified by a widespread network of 19 U.S. gardens working with the Center for Plant Conservation (CPC). The CPC estimates that 3,000 taxa native to the United States are threatened with extinction; of these, more than 300 are now cultivated in the network of gardens (R. Stafford, CPC, personal communication, May 1989).

The contribution of botanic gardens to the conservation of species extends beyond the preservation of species threatened in the wild.

Botanic gardens supply plants for research and horticulture, thereby taking pressure off wild populations, and they are important educational resources. Each year, an estimated 150 million people visit botanic gardens (IUCN/WWF 1989).

The already important role of botanic gardens could easily be expanded. To begin, the current geographical imbalance in the locations of botanic gardens could be remedied if more gardens were established in tropical countries. Today, tropical countries possess only 230 of the world's 1,500 botanic gardens (Heywood 1987). While over 100 new gardens have been opened or planned in the last decade and many of these are in tropical regions, the geographic imbalance persists, particularly considering the greater species richness of tropical regions.

Second, with further research into storage techniques and with better data on where specimens were collected (''passport information'') and their breeding history, botanic gardens could become even more important genetic repositories. The IUCN Botanic Gardens Conservation Secretariat is now developing a computerized data base of species occurrences in botanic gardens to help gardens collect species that are absent or underrepresented in captivity. The efforts of botanic gardens in germplasm conservation are being coordinated by the IUCN Botanic Garden Conservation Strategy; in association with the IBPGR, IUCN is also putting together guidelines for collecting germplasm of wild species.

Until recently, botanic gardens have been underutilized in the maintenance of threatened species and the conservation of genetic resources. Even though they contain a large proportion of the world's flora, the gardens have traditionally not been integrated and their holdings have been known only locally. Thanks to the efforts of numerous committed individuals and institutions, their role in conservation is growing rapidly.

VII. Summary and Recommendations

At a time when humanity's needs for productive biological resources are greater than ever before, we are witnessing the irreplaceable loss of the world's fundamental capital stock—its species and genes—and the deterioration of ecosystems' ability to meet human needs. As species disappear, humanity loses today's foods, medicines, and industrial products, as well as tomorrow's. As genetic diversity erodes, our capacity to maintain and enhance agricultural, forest, and livestock productivity decreases. And with the degradation of ecosystems, we lose the valuable services that natural and semi-natural systems provide. Maintaining the potential of this "capital stock" to meet human needs must be a priority as we pursue sustainable development. But each day that these losses continue, the pursuit becomes increasingly difficult.

If we don't act immediately, extinctions in the coming decades may represent the most massive loss of species since the end of the Cretaceous Era, some 65 million years ago. Already, the rate of extinction of birds and mammals may be as much as 100 to 1000 times the background extinction rate. The single greatest cause of species extinction in the next half-century will be tropical deforestation. Scientists concur that roughly 5 to 10 percent of closed tropical forest species will become extinct *per decade* at current rates of tropical forest loss and disturbance. With more than 50 percent of species occurring in closed tropical forests and a total of roughly 10 million species

on earth, this amounts to the phenomenal extinction rate of more than 100 species per day.

The extinction crisis is not restricted to tropical forests. Freshwater habitats are being dramatically altered as rivers are impounded and exotic species introduced. In the southeastern United States, for instance, 40 to 50 percent of freshwater snails have been driven to or near extinction as water impoundments were built and rivers straightened, widened, and deepened. Oceanic islands, where most extinctions have occurred in past centuries, also remain among the most threatened ecosystems on earth. Some 60 percent of the plant species endemic to the Galápagos islands are threatened with extinction, as are 40 percent of Hawaii's endemic species and 75 percent of the endemic plant species of the Canary Islands. Mediterranean climate zones, with their high plant species richness and distinctive floras, face continuing threats of species loss through habitat conversion and species introductions.

Habitat loss and degradation are the most important causes of the extinction crisis, but overharvesting, species introductions, pollution, and other causes also take a significant toll. Global warming will exacerbate the loss and degradation of biodiversity by increasing the rate of species extinction, changing population sizes and species distributions, modifying the composition of habitats and ecosystems, and altering their geographical extent. Even if all human impacts on the biosphere were to cease immediately, species extinctions due to the impacts that have already taken place would continue for decades.

Genetic diversity, an invaluable tool used by farmers since the dawn of agriculture, faces extinction threats just as great as those to species. Wild relatives of economically important crops, trees, and livestock—often carrying unique genes that can be used to breed resistance to pests or to improve yields—are disappearing, and crop cultivars that farmers have planted for centuries are being replaced with high-yielding varieties. Seed and germplasm banks have been established and expanded in the past two decades to deal with the threats to genetic diversity, but significant problems and gaps remain. For example, nearly half of the two million accessions currently stored in gene banks have not been described. Moreover, while crops of global importance—such as rice and wheat—are well represented in seed banks, such regionally important crops as cassava are not.

This report has shown how the basic components and processes of the biosphere are linked to each other and to the goods and services they provide to humanity. These multiple interconnections make it clear that species can't be managed without managing their genetic diversity and the habitats in which they occur and that optimal conservation priorities can't be set unless the connections between biodiversity and human needs are understood. Much of biodiversity is used directly—as, say, the basis for breeding programs or as foods and medicines—but such ecological processes as production, decomposition, nutrient and water cycling, soil generation, erosion control, pest control, and climate regulation are also essential to human survival.

Considering the wide range of services that ecosystems provide, ''biodiversity conservation'' cannot be considered synonymous with ''endangered species preservation.'' Endangered species preservation is an essential damage-control activity, one component of a strategy to conserve biodiversity and a useful yardstick for judging success in conserving biodiversity. But biodiversity conservation also focuses on species not yet threatened with extinction and on the habitats and assemblages they comprise. It is far easier to protect species through foresight while populations are still large than it is to attempt to pull an endangered species back from the brink of extinction, and species conservation is best achieved through conserving habitats and ecosystems—not through heroic rescues of individual species.

A variety of tools, including both *in situ* and *ex situ* techniques, play critical roles in the conservation of biodiversity. *In situ* preservation of genes, species, and habitats ensures the maintenance of areas in a natural or semi-natural state. Protected areas—ranging from those where human use is minimized to areas designated for, say, timber harvest—can both maintain biodiversity and provide jobs and resource-based products. Dedicating a relatively small portion of the earth's land surface to protected areas can save a substantial fraction of the world's species diversity. *In situ* conservation of agricultural genetic diversity ensures the maintenance of both the genetic material and the farming system with which it has evolved and allows crop and livestock varieties to continue to adapt to such changing environmental conditions as increases in pest populations or temperature.

Besides contributing to public education and basic research, *ex situ* conservation complements *in situ* techniques and must be closely coordinated with *in situ* conservation in both natural and human-altered ecosystems. Zoos, aquaria, and botanic gardens are already playing significant roles in biodiversity conservation and their utility can be enhanced considerably with further research and greater financial commitment to their activities. *Ex situ* preservation of agricultural genetic diversity through seed banks and other techniques affords breeders ready access to a wide range of genetic materials and may be a precious last resort for many species and varieties that would otherwise become extinct as modern varieties replace them. The utility of *ex situ* conservation of genetic resources can be strengthened considerably by placing greater emphasis on conserving and maintaining the genetic diversity of crops important in tropical countries, the wild relatives of economically important species (particularly using *in situ* techniques), the genetic diversity of important forestry species, livestock's genetic diversity, and knowledge of the farming systems in which local crop and livestock varieties evolved.

During the next few decades, the world will experience the largest growth of human population in history. Increasing population, and even more rapidly increasing demands for biological resources, will strain the world's remaining biological diversity. With growing pressures on biological resources, and with ecosystem degradation increasingly hindering people's ability to meet daily survival needs, the restoration and rehabilitation of ecosystems will become a major conservation focus in coming decades. An important rationale for conserving biodiversity is to maintain species of potential value for their use in restoring and rehabilitating degraded ecosystems—of particular concern given global warming's probable impacts on biological diversity.

Considering the gravity of the threat to the world's biodiversity and the toll that the degradation of biological resources is exacting on individuals and development efforts throughout the world, quick action is needed to halt the loss of biodiversity. Such action must address many of the root political, social, and economic causes of the loss and degradation of biological diversity, but it must also be informed by the biological sciences.

The following recommendations for addressing the biodiversity crisis reflect this all-important biological perspective. They also complement proposals put forth elsewhere for increasing the developed countries' financial commitment to biodiversity conservation. The International Conservation Financing Project of the World Resources Institute suggested that unmet conservation financing needs in developing countries total as much as $20 billion to $50 billion per year (1 to 2 percent of the Third World's GNP) (WRI 1989). The cost to all countries of *not addressing* the biodiversity crisis is, of course, far greater than this estimate, but it will not appear on current national income accounts. Instead, the price will be paid by future generations.

The response of developing countries to the threat of biological impoverishment is particularly constrained by pressing social and economic issues and by the burden of international debt. Many actions that can be taken to stem the loss of biodiversity do provide short-term economic benefits—say, maintaining natural forests so that wild species can be harvested for food, medicines, and industrial products or establishing protected areas so that tourists will visit, or downstream water development projects won't clog with silt. But the long-term conservation actions needed will not be taken in the developing world unless developed countries provide their fair share of the funds, technology, and knowledge needed to implement them. All countries share responsibility for conserving biodiversity, and maintaining biodiversity is in the interest of every nation and in the interest of the global community.

1. Enhance the Foundation for Decision-making

Scientists now know enough about the distribution of biological diversity, the threats that it faces, and the conservation tools available to maintain it so that current conservation efforts could be expanded considerably without fear of wasting effort or money. But important remaining gaps in knowledge will continue to hinder conservation and limit the benefits that biological resources can provide to humanity unless research programs are greatly strengthened. Indeed, as understanding of biodiversity grows, so does recognition of the pressing needs for further research.

The current financial commitment to bio-diversity research is extremely low. Only $16.7 million was spent by U.S.-based organizations (governmental and non-governmental, including universities and museums) on biological diversity research activities in developing countries in 1987 (Abramovitz, in press). This amounts to only 0.2 percent of all foreign assistance provided by U.S. AID in that year. To make matters worse, only a small number of qualified scientists are capable of conducting such research projects. No more than 1,500 or so professional systematists worldwide have been trained to carry out taxonomic research on tropical species (NRC 1980).

Against this backdrop, the following research is particularly urgent:

a. *Accelerate inventory and systematics research.* The number of experts in the taxonomy and distribution of the world's biodiversity must be increased substantially, partly so that biological inventories can be accelerated. To improve the foundation for establishing conservation priorities, emphasis should be placed on inventorying lesser-known taxa. In addition, more rapid means of predicting regions of high diversity or endemism are needed since basic conservation priorities will have to be set in the absence of detailed inventories. For example, soil, climate, and topographical information alone can be used to roughly delimit regions of probable value for bio-diversity conservation. This first-cut analysis can then be supplemented by information on plant or vertebrate diversity. Particularly in the tropics, scientists do not yet know how well this streamlined approach will conserve other groups of organisms, such as insects. Comprehensive studies of all taxa at several research sites would begin to answer this important question.

Until species are at least named and described, whatever properties and traits of potential use to humanity they might possess can't be assessed. Thus, needs for increased taxonomic research go hand-in-hand with needs for increased screening of species for their potential value as medicines, foods, or industrial products. Given the small number of taxonomists available, the relatively high cost of this type of research if undertaken by international organizations, and the size of the challenge, high priority should be given to enhancing the research and screening capacity of local institutions in developing countries. Since species inventory and initial screening at the local level are labor-intensive and inexpensive, training individuals in developing countries to undertake this work will help build strong in-country capacity for studying and conserving biological diversity and create sorely needed jobs.

b. *Increase support for community and ecosystem ecology research.* Accelerated research is essential in four areas of community and ecosystem ecology:

• The relationship between biodiversity and ecological processes is poorly understood. Further research on water and nutrient cycling, ecological energetics, ecosystem stability, soil formation, and other processes will help scientists predict the consequences of extinctions or the impact of changes in land-use on ecological processes more accurately. Research on communities and ecosystems will help them find out which species are functionally redundant and which are keystone species whose loss is likely to reverberate throughout an ecosystem. Management strategies for biological resources could also benefit greatly from such work.

• Research on the effects of fragmentation on biological diversity and ecological processes must be accelerated. Greater understanding of how fragmentation affects natural and semi-natural ecosystems will improve the design of protected areas, help scientists develop management plans for individual species, and improve predictions of fragmentation's impacts on various ecological processes. Such research can also help conservationists identify species at risk of extinction due to fragmentation and manage habitat fragments in a "supersaturated" state—with, that is, greater species richness than the area of the habitat would naturally allow. Further studies of the ecology of ecotones, edge effects, the effects of fragmentation on dispersal, and the utility of conservation corridors are also needed to aid in the design and management of protected areas.

- Community restoration and rehabilitation of degraded lands are quickly becoming two of the most important tools for conserving biodiversity and for meeting people's needs for productive biological resources. Both techniques require greater knowledge than scientists now have of the ecology of the community being restored, and rehabilitation requires further knowledge of exotic species suitable for introduction. The potential impacts of climate change on ecological systems add further urgency to the need for this research. More particularly, the ability to predict the composition of communities following climate change must be enhanced so that restoration techniques can be used in anticipation of new climatic conditions, and new techniques are needed to help establish species under climatic conditions that are expected to become favorable to the species within several decades.

- The monitoring of biodiversity and ecological processes, essential for decision-making, needs far greater research emphasis. Monitoring schemes, as well as sampling procedures for ecosystem and population data, must be redesigned to give the maximum information useful to decision-makers at a minimum cost. Studies must be directed at easily monitored bellwether species that can reveal changes in the entire community or ecosystem. Toward these ends, greater use of remote-sensing techniques for monitoring should be encouraged in all countries.

c. *Boost research in population ecology.* The maintenance of biodiversity requires an understanding of the factors influencing the persistence of populations of individual species. The approach known as "population viability analysis," which attempts to determine the minimum viable population of species, is proving extremely valuable in assessing land-management options, management plans for single species, protected area designs, and *ex situ* breeding programs. Further research into the demographic and genetic factors influencing small populations and support for the development of better models for population viability analysis are vital.

d. *Integrate the study of cultural diversity into biodiversity research.* Biodiversity conservation has traditionally focused on species and genes and ignored the human culture with which biodiversity coevolved. Now, knowledge of human farming and resource use systems must be increased before it is irretrievably lost. Further, studies of traditional methods of resource management can provide important insights into techniques for sustainable resource management in tropical regions (Oldfield and Alcorn 1987, Brush 1989).

Far from comprehensive, this list of particularly important research needs indicates the breadth of research needed to lay a more firm foundation for conservation decision-making. Although the focus here is on *in situ* conservation, needs for research on *ex situ* conservation—including the study of seed, sperm, and embryo-storage techniques, captive propagation, the genetic characterization of populations, and techniques for reintroducing species into the wild—are pressing too. Since a major goal of biodiversity conservation is to increase the benefits based in the world's living resources, in the decades ahead identifying and developing new foods, medicines, and industrial products derived from species is also extremely important. (More detailed reviews of research needs for the conservation of biodiversity are provided in Wildt and Seal (1988) and Soulé and Kohm (1989).)

2. Establish Biodiversity Conservation as a Development Priority

While it is clear that long-term needs for development require the conservation of biodiversity, people's short-term needs often lead to pressures on natural and semi-natural systems that accelerate its loss. Biodiversity conservation can't attain priority status among development objectives until the conflicting pressures of short-term and long-term human needs are reconciled. Following are the two most promising avenues toward this goal:

a. *Expand the short-term utility of biodiversity.* Increasing the flow of benefits that people derive from the world's biodiversity both strengthens the rationale for its short-term conservation and solves other pressing social and economic problems. Convincing demonstrations of the potential for developing new products, medicines, or foods from wild species and the potential for development projects linking protected-area conservation with rural development are needed to realize this goal. Such models would help build developing countries' national capacity for research and development. So would research aimed at defining the national and regional research and development priorities and the institutional requirements for developing new products and services that contribute to an ecologically sustainable future, generate new sources of employment, and help local economies use local resources to become more self-sufficient. Specifically,

germplasm banks, enhanced inventories of basic species, crop and livestock breeding facilities, medicinal screening facilities, agricultural research, and so forth can all boost biodiversity's contribution to human welfare.

b. *Affix true economic values to biodiversity.* A key to making biological resource conservation a development priority is accurately assessing the economic values of biodiversity and the costs of degrading or mismanaging biological resources. The values of biodiversity fit poorly into mainstream economic evaluations. Establishing ''existence values'' (the value of knowing that a species exists), exploring option values (the potential future value of a species of no current economic importance), and affixing value to irreplaceable items are notoriously difficult undertakings, yet all are central to the economic valuation of biodiversity. Many of the immediate values of biodiversity—among them, the supply of food, fuel, and medicines to rural people—are also difficult to establish because few of these goods are traded outside of local markets. It's also extremely difficult to assign values to biological diversity's role in ecosystem functioning. If biodiversity conservation is undervalued relative to such activities as logging or agriculture, it will be short-shrifted when national governments and multilateral and bilateral aid agencies make land-use decisions. To right this situation, resource economists must evaluate both the direct and indirect values of species, and they must lower the discount rates they use when assessing irreplaceable resources.

3. Encourage Integrated Regional Planning and Cross-Sectoral Coordination

Institutions in most countries are ill-equipped to handle biodiversity conservation because they are organized along sectoral lines and biodiversity conservation requires a broader view. Responsibility for managing the world's biodiversity has been fragmented into ministries of agriculture, defense, education, fisheries, forestry, health, and parks. From a biological standpoint, the need for an integrated regional approach to conservation is readily apparent. Almost all ecosystems are ''open,'' in the sense that they exchange energy, mineral nutrients, and organisms with other ecosystems. Consequently, biosphere management systems must be open too (Noss 1983). Strategies for conserving the world's biological diversity cannot succeed if each patch of natural and semi-natural land is managed in isolation, or if these lands are conceptually isolated from planning and economic development.

Institutional fragmentation undermines each sector's effectiveness in meeting biodiversity-conservation objectives. When range managers, agricultural experts, foresters, and fisheries experts make land-use decisions along sectoral lines, the multiple ecological effects of decisions are often overlooked in the attempt to maximize one form of production—be it head of cattle, bushels of wheat, board-feet of wood, or tons of fish. To end this Balkanization, institutional responsibilities for land-use planning and management must be distributed so that trade-offs involving more than one sector are evaluated from more than one angle, and close cooperation with the rural people who in most cases manage the world's biological resources is essential. The regional integration and coordination of resource planning and management activities helps ensure that lands or seas managed for various objectives meet people's immediate needs while maintaining the maximum possible biodiversity.

Following are two important steps toward achieving the inter-sectoral integration necessary for biodiversity conservation:

a. *Explore innovative institutional management arrangements.* The potential for improving the capacity to maintain genes, species, and habitats through integrated planning and inter-sectoral coordination is great, but

such coordination has often proven difficult to achieve. Still, recent initiatives in Costa Rica have successfully integrated previously separate land-management activities, partly to maintain biological diversity. A new program for land management places the various parks, forests and other reserves—which together make up 27 percent of Costa Rica's area—under seven regional authorities that report to the Minister of Mines and Natural Resources. All agricultural, forest, and grazing lands, and villages and towns found within each region are under the jurisdiction of each regional unit, and these units work both with protected area management and with community-oriented development programs in which conservation figures centrally. If these so-called mega-reserves are properly managed, more than 95 percent or so of the species that existed in Costa Rica before Europeans arrived could be maintained (D. Janzen, University of Pennsylvania, personal communication, Dec. 1988).

b. *Establish appropriate management incentives.* Developing appropriate conservation incentives for all managers of land or coastal areas is one key to encouraging inter-sectoral coordination. Over the long term, such sectors as forestry, fisheries, agriculture, tourism, and watershed management will all benefit from biodiversity conservation, but short-term incentives to increase timber harvest, fishery yield, crop production, or tourism revenues often work against conservation efforts. Incentives for managers must therefore encourage them to increase the emphasis on biodiversity conservation while still meeting other resource use goals and to coordinate land-use and management objectives inter-sectorally. In particular, managers need incentives to develop management techniques that maximize the conservation of biodiversity in semi-natural lands dedicated mainly to meeting other land-use objectives.

4. Establish a Strategy to Guide Biodiversity Conservation Activities and Priorities

Each nation and institution has its own bio-diversity-conservation perspectives, values, and goals, and no single set of priorities for biodiversity conservation will ever be universally accepted. Trade-offs and value judgments concerning the distinctiveness of species, habitats, or ecosystems, their current or future utility, their local or global importance, and the severity of threats to biodiversity will influence strategies for halting biological impoverishment. *(See Box 6.)* Nevertheless, at all levels of action—from the grassroots to international organizations—coordinating the various conservation activities will enhance biodiversity conservation. Cooperative efforts to identify conservation priorities can maximize the amount of information available to decision-makers, the coverage of conservation actions, and financial support for conservation. The key to such cooperation is identifying priorities for action, the roles of various institutions, financial requirements, sources of financing, opportunities for policy or institutional reform, public education requirements, training needs for resource managers, policy analysts, and resource economists, and so forth. A cooperative strategy or action plan generally works best when all institutions and individuals involved in and affected by resource management have a hand in shaping its content and implementing it.

One of the most important benefits of a widely supported global, national, or local strategy for conserving biodiversity is that it can ensure that immediate conservation needs are met while long-term work on many of the root causes of biological resources degradation continues. It can also ensure that research and action are not viewed as competitors for scarce resources, but rather as equally essential components of conservation—both of which deserve increased support.

The specific approach taken by various nations to develop biodiversity conservation strategies is likely to vary from the establishment of a National Biodiversity Action Plan to more general Environmental Action Plans or National Conservation Strategies to local programs addressing such specific issues as coastal zone management. To encourage such planning at all levels and to help expand the conservation constituency, identify opportunities for addressing the root causes of the loss of bio-diversity, encourage the best use of modern science, and help elevate biodiversity conservation on the list of national priorities, WRI, IUCN, and UNEP are seeking to organize the development of a Global Strategy for Conserving Biodiversity (McNeely et al. 1989, Miller et al. 1989). The program will be committed to three inextricably linked pursuits: *saving* the maximum diversity of life, *studying* its potential use and its role in the biosphere, and *using* it in sustainable management systems. The Global Strategy, to be developed over two years, would be implemented through a decade of intensive and sustained world-wide effort by institutions and individuals.

5. Promote Biocultural Conservation

Humanity has long caused extinctions, altered succession, and modified the population size of other species. It has domesticated species and even altered the genetic diversity of wild populations. These historic manipulations of the environment were probably no "wiser" than our current modifications; many, in fact, led to environmental catastrophes, as the extinction of Hawaii's avifauna *(Box 3)* or the desertification of the Indus River Valley attest (Bryson and Murray 1977). Nevertheless, human culture's imprint on the world's bio-diversity and the close connections between traditional resource management systems and local biological diversity should not be overlooked since the maintenance of "natural" ecosystems may sometimes require the continuation of historic human disturbances. Then too, many of the resource management systems that human societies have developed represent unique and often extremely useful methods of responding to local environmental conditions—potential models for further adaptation.

Box 6. Criteria for Setting Biodiversity Conservation Priorities

Given that some species and genes are sure to be lost and some natural habitats converted to other land uses in the coming decades, how should scarce human and financial resources be spent to minimize biotic impoverishment and maximize biodiversity's contribution to human well-being? How can the most useful components of biodiversity be saved for current or future use? And how can such ethical considerations as our responsibility to other species and to future generations be incorporated into conservation priorities?

Clearly, these questions cannot be answered with scientific information on the nature, distribution, and status of biodiversity alone. Such factors as awareness of the problem and its solutions, perceived trade-offs with other basic human needs, and local political considerations will all influence conservation priorities. Nonetheless, three rules of thumb derived from the biosciences can help scientists and policy-makers identify the various influential factors and evaluate the trade-offs and value judgments made in setting priorities.

1. Distinctiveness: Numbers Aren't Everything.

To conserve the widest variety of the world's life forms and the complexes in which they occur, the more distinctive elements of this diversity should receive high priority. For example, preserving an assemblage of plants and animals is comparatively more important if those species are found nowhere else or are not included in protected areas. Similarly, if the choice is between conserving a species with many close relatives or a species with few, the more distinctive species should be saved.

Maintaining the highest number of species without considering their taxonomic position makes little sense, as a comparison of marine and terrestrial environments shows. Terrestrial environments contain at least 80 percent of the world's total species, mainly because vascular plants and insects are so numerous on land—accounting for nearly 72 percent of all described species in the world—and so poorly represented in marine environments. However, the sea contains greater proportions of higher taxonomic units. Marine ecosystems contain representatives of some 43 phyla while terrestrial environments are home to only 28 phyla. The sea contains fully 90 percent of all classes and phyla of animals (Ray 1988). Clearly, efforts to conserve the widest array of biodiversity must give attention to all levels of the taxonomic hierarchy.

2. Utility: Global or Local, Current or Future?

To some extent, utility is always a criterion in setting biodiversity conservation priorities. Nobody opposed the extinction of wild populations of the smallpox virus, and most people would agree that it's more important to conserve a subspecies of rice than a subspecies of a "weed." Similarly, maintaining forest cover in the watershed above a village or irrigation project seems more important than preserving a similar forest in a region where increased flooding or soil erosion would not affect human activities downstream.

Of course, many decisions about which species to save are much more difficult than the one governing smallpox's fate. In general, questions about the usefulness of species, genes, or ecosystems can be complicated by two factors. One is the issue of inter-generational equity—whether today's needs outweigh those of future generations. It seems reasonable enough to expect future

Box 6. (cont.)

generations to value the species most valued today, but it is impossible to predict humanity's exact needs, and many other species of unknown value today are sure to become more important in the future. Amid attempts to conserve species prized today, some value has to be assigned to future generations' needs.

━━━━━━━━━━━━━━━━━━

How useful species, genes, or ecosystems are depends on whether today's needs outweigh those of future generations and who gets the benefits.

━━━━━━━━━━━━━━━━━━

Second, we must ask the question: "Utility for whom?" Biodiversity has obvious values to local communities, nations, and the entire world, but the benefit is not identical for each group. From a global perspective, the conservation of a region's biodiversity might help to regulate climate, influence the atmosphere's chemical composition, and provide all of humanity with industrial products, medicines, and a source of genes for crop breeding. Locally, conservation may also provide people with fuel, clean water, game, timber, aesthetic satisfaction, and important cultural symbols or resources. Because the conservation benefits received by the local and global communities are not congruent, international and local priorities will also differ. To humanity at large, conserving tropical forests matters more than conserving semi-arid deserts since the forests contain a tremendous variety of life and heavily influence global climate. Locally,

however, each region's biodiversity is equally valuable since it provides essential ecosystem services that local people rely upon. Neither perspective is necessarily the "correct" view of biodiversity; either—global or local, current or future—reflects an implicit value judgment.

3. Threat: Saving the Most Beleaguered Species and Ecosystems First.

In general, the threat to biological diversity is influenced by how widespread species are, how common they are over their range, and by such human pressures as harvesting, land conversion, and environmental pollution. The threats to species vary dramatically by region, and in each region some ecosystems are more threatened than others. For instance, Central America's tropical rain forests are less threatened than the remaining fragments of tropical dry forest in that region. In such cases, the dry forests should receive the most attention even though the rain forests contain more species.

These three factors—distinctiveness, utility, and threat—help make value judgments more explicit as conservation priorities are set. If a strictly global perspective is adopted, then preserving the most species-rich sites in the world seems most important. Indeed, this global perspective has led some organizations to focus on conservation in the "megadiversity" countries—Brazil, Colombia, Indonesia, Madagascar, Mexico, Zaire, and others *(see discussion in McNeely et al. 1989)*. In contrast, from a nationalist perspective, each country's biodiversity is worth roughly what every other country's is. Similarly, if today's needs for genes, species, or ecosystems are considered paramount, tomorrow's needs can't be too.

To maintain some "natural" ecosystems, we might have to continue disturbing them.

The following actions will help promote a *biocultural* approach to conservation:

a. *Ensure that human culture and knowledge is conserved as part of conserving biodiversity.* Fraught with both success and failure, the history of human management of the environment contains important hard-bought lessons that could easily be lost. While 20th-century western civilization congratulates itself for the green revolution, these "revolutionary" gains pale in comparison to the enhanced productivity that resulted from the slow process of domestication that first turned weeds into productive crops. The seeds of domesticated crops that we seek to preserve in germplasm banks are only one part of an agricultural technology developed as farming systems and their crops coevolved. When these farming systems and seeds face extinction, *ex situ* storage of germplasm is essential and so is the preservation of agricultural knowledge.

b. *Enhance traditional systems of resource use, where appropriate, rather than replacing them.* Indigenous systems of land management are often technologically and institutionally well adapted to local environmental conditions. In southern Para in Brazil, the Kayapo people are practicing traditional methods of soil enrichment and cropping that allow them to use cleared plots for eleven years with as little as a five-year fallow period (S. Hecht, University of California, Los Angeles, personal communication; May 1988). In contrast, most colonists in the Amazon who have not had long experience with farming find cleared plots unusable after two to three years.

Although many traditional community-based resource management systems have succeeded, efforts to establish them where no historical precedent existed have frequently suffered from the "tragedy of the commons." In Nepal, community forests were once successfully protected by forest guards with the power to prevent cutting of protected areas, determine where trees could be cut, inspect firewood stocks in people's houses, and levy appropriate fines (von Fürer-Haimendorf 1964). Modifying these traditional systems may make far more sense than replacing them (McNeely 1989). Precedents for reviving age-old practices do exist: local villagers were enticed to return to traditional methods of wildlife conservation in the Lupande Game Management Area of the Luangwa Valley when a Wildlife Conservation Revolving Fund was created. Revenue generated from safari fees and the sale of meat and hides goes into the fund and supports both wildlife management and local community development (Lewis et al. 1987).

6. Anticipate the Impacts of Global Warming

Although global warming will exacerbate biotic impoverishment, society's response to these impacts is likely to be severely constrained by other pressing social needs that will arise if climate changes rapidly. Because needs to mitigate impacts on agriculture, water resources, and coastal development are sure to take precedence over steps to mitigate impacts on biological diversity, the most effective way to minimize the impacts of climate change on biodiversity is to slow or halt climate change.

Of course, little is gained if actions to slow climate change are detrimental to biodiversity. For example, one strategy for slowing carbon-dioxide buildup is to increase forests' carbon sequestering role by replacing natural forests with plantations of fast-growing trees—a response that will take a significant toll on biological diversity. A strategy more compatible with the need to minimize negative impacts on biological diversity would be to adopt agroforestry schemes, which can help increase carbon-dioxide uptake and offer people a ready supply of wood products, thereby lessening demands for wood and land from natural forests.

Uncertainty about climate change's specific effects on biodiversity will persist, as will uncertainty about the nature and extent of climate change. For this reason, the most desirable policies for mitigating its impacts are those that would work under any probable scenario of climate change and ecosystem response. Fortunately, many current biodiversity conservation efforts meet this criterion, including efforts to strengthen and enlarge protected areas (especially with poleward or up-slope additions), establish conservation corridors, enhance *ex situ* conservation capacity, provide benefits from protected areas to people in surrounding communities, increase research on species taxonomy, distribution, ecology, and conservation biology, and increase agricultural productivity.

Apart from slowing the rate of climate change and enhancing ongoing conservation efforts, the measures that can be taken to mitigate effects on biodiversity include the following:

a. *Increase land-use flexibility.* As global climate changes, many species distributions will change. Current systems of land-use and land tenure make no allowance for these distributional shifts; they need to be adjusted so that options for converting land to natural or semi-natural states are available should such conversion be needed. For example, the potential effects of global warming on coastal wetlands could be exacerbated if upland sites that wetland species could colonize as sea level rises aren't available. Similarly, land ownership changes will sometimes be required to re-create plant and animal communities upslope or poleward of existing assemblages. In the case of wetlands, it makes sense to adjust the boundaries of protected areas based on predicted sea-level contours by purchasing development rights for inland areas and restricting coastal development.

b. *Enhance response techniques.* Both *ex situ* preservation and ecological restoration will be needed to mitigate the impacts of greenhouse warming on biodiversity. But policymakers must recognize that neither of these technologies will be a panacea. *Ex situ* preservation cannot conserve habitats, ecosystems, or ecological processes, nor can it allow the continued evolution of species. Ecological restoration is expensive, particularly if undertaken on a large scale, and chancy since scientists cannot precisely predict the community composition that is likely under future climatic regimes. Site-specific predictions won't be reliable enough to guide investment, so most conservation responses to climate change must be reactive rather than anticipatory. Nevertheless, both of these techniques will become increasingly important under conditions of rapid climate change, and greater research and investment is needed now to improve them.

Walter V. Reid is an Associate with the WRI Program in Forests and Biodiversity. Before joining WRI, Dr. Reid was a Gilbert White Fellow with Resources for the Future. He has also taught at the University of Washington and worked as a wildlife biologist with the California and Alaska departments of Fish and Game and with the U.S. Forest Service in Alaska. **Kenton R. Miller** is Director of WRI's Program in Forests and Biodiversity. Prior to joining WRI, Dr. Miller served for five years as Director General of IUCN. Previously, he directed the Center for Strategic Wildland Management Studies at the University of Michigan and, before that, the U.N. Food and Agriculture Organization's wildlands management activities in Latin America and the Caribbean.

Appendices

Appendix 1: Cascade Effects

The loss of a species can have various effects on the remaining species in an ecosystem—what kind and how many depends upon the characteristics of the ecosystem and upon the species' role in its structure. Cascade effects occur when the local extinction of one species significantly changes the population sizes of other species, potentially leading to other extirpations. Such cascade effects are particularly likely when the lost species is a "keystone predator," a "keystone mutualist," or the prey of a "specialist predator."

Keystone Predators

How much the loss of a predator affects its prey population size depends on how much the predator limits the prey population. If the size of the prey population is determined by factors other than predation, it is said to be *donor-controlled*. For example, such bottom-dwelling invertebrates as mussels or barnacles consume only a small fraction of the ocean's plankton; consequently, removing these invertebrates would not affect the plankton's population size much. Similarly, although the passenger pigeon *(Ectopistes migratorius)*, Carolina parakeet *(Conuropsis carolinensis)*, and wild turkey *(Meleagris gallopavo)* were the dominant large seed-predators of eastern North America's forests, the extinction of the first two species and the near-extinction of the third apparently changed the forest little. The loss of the bird species increased seed survival, but this increase had no effect on the density of trees because what limited the tree populations was the availability of space, not seed predation (Orians and Kunin, in press).

In a *predator-controlled* system, the size of the prey population is determined by predation. In such systems, the impact of the loss of the predator can be substantial, especially if the predator is relatively high in the food chain. So-called keystone predators affect not only their prey's population size but also the community's species diversity. By limiting the population size of a species that would otherwise outcompete other species, keystone predators can help maintain high species diversity. When the keystone predator *Pisaster ochraceus* (a starfish) is removed from the intertidal zone in the Pacific northwest of the United States, the intertidal community changes from one with a high diversity of relatively large bottom-dwelling invertebrates to a virtual monoculture of the starfish's favorite prey—the mussel *(Mytelus edulis)* (Paine 1966).

Keystone predators represent a vexing problem for conservation. Many keystone predators, such as sea otters, face a high risk of extinction because they tend to be high in the food chain and relatively sparsely distributed. *(See Box 4.)* Yet, determining which species play keystone roles in a community may be extremely difficult.

For example, the small moth *Cactoblastis cactorum*, introduced into Queensland from South America, appears to be a minor component of the Queensland community since its population size and biomass aren't great. However, it is a keystone predator in Australia and probably also in South America. In the absence of the moth, *Opuntia* cactus (also introduced from South America) covered 62,000 square kilometers of Queensland but in its presence *Opuntia* has been reduced to small isolated patches (DeBach 1974). In general, determining whether a given species plays a keystone role requires considerable study and experimentation.

Although certain species have much more influence than others on an ecosystem's structure, not all ecosystems include a single species that exerts such a pervasive influence. In fact, most ecosystems are somewhat sensitive to the loss of any one of many species, though some losses have greater impact on the system than others. Nevertheless, determining which species function as keystone predators in an ecosystem can help managers fine-tune conservation activities.

Mutualists and Keystone Mutualists

Species involved in mutualisms—mutually beneficial interactions such as pollination—are deeply affected if their partner mutualists are lost. For example, the loss of a fruit-eating species may severely affect the plants whose fruits it disperses. Similarly, rare tropical plants dependent on specific species for pollination would be threatened with extinction by the loss of the pollinator (Orians and Kunin, in press). Some orchids, for instance, rely on a single species of euglossine bee for pollination; without the bee, the plant would become extinct (Dodson 1975, Gilbert 1980). By the same token, pollinators that specialize on a specific plant species could die out if the plant does. Most species of fig (*Ficus* spp.) are each pollinated by a different species of wasp, and each species of wasp pollinates only one fig (Janzen 1979), so the loss of a fig species could wipe out one of the specialist wasps.

Maintaining plant-pollinator interactions in tropical forests is of utmost importance since tropical forests are much more sensitive to extinctions of pollinators than are temperate forests. Whereas less than 30 percent of trees in the northern United States are pollinated by animals (as opposed to wind), more than 95 percent of the trees in one dry forest in Costa Rica are believed to rely on animal pollination (Regal 1982).

Certain species—keystone mutualists—involved in mutualistic interactions may sometimes assume great importance in the community. For example, during the dry season in the tropical forest at the Cocha Cashu Biological Station in Manu National Park of southeastern Peru, only 12 of approximately 2,000 plant species support the entire fruit-eating community of mammals and birds (Terborgh 1986a, 1986b). During this period, fruit production drops to less than 5 percent of the peak production, so the food available to the fruit-eating species is severely restricted. The 12 plant species still producing fruit and nectar meet the food needs of as much as 80 percent of the entire mammal community and a major fraction of the avian species at the site. Clearly, the loss of one or more of these ''keystone mutualists'' could significantly harm frugivore populations.

While the identification of keystone predators may be extremely difficult, keystone mutualists can generally be identified through non-experimental observations of the resources that the species use. Once identified, keystone mutualists can be managed to ensure the survival of the species dependent upon them. By increasing the abundance of the keystone mutualists for instance, it might be possible to increase populations of dependent species (Terborgh 1986a).

Specialist Predators

In general, the loss of a species tends to have greater impacts on its own predators if the species is relatively high in the food chain. Populations of such species, including carnivores, tend to be limited by the amount of prey available, whereas species low in the food chain, such as herbivores, are more often limited by their predators or pathogens (Hairston et al. 1960, Oksanen 1988). Consequently, the loss of an herbivore tends to directly reduce the supply of the factor (food) limiting the carnivore population, while the removal of a plant species may have no effect on the herbivore population.

One exception to this pattern occurs when the plant species removed is the prey of a specialist predator—that is, a species with no alternative food source. For example, many insects eat just one species of plant because they have evolved mechanisms for detoxifying the plant's chemical defenses. If such a plant were lost, the highly specialized insect would be too. Widespread and common prey species are more likely to have specialist predators than are rare or locally endemic species (Strong 1979, Orians and Kunin, in press), so the loss of common species would tend to have a large impact on species higher in the food chain.

Appendix 2: Calculating Extinctions Due to Deforestation

The loss of habitat has predictable effects on the number of species in a region. Larger areas of habitat tend to contain more species. With the use of a species-area curve representing this relationship, it is possible to predict the proportion of species that will become extinct in a region based upon the amount of habitat that is lost. (See Figure 7.) To make such predictions, it is necessary to estimate the rate of habitat loss and to choose a species-area curve appropriate for the habitat in question.

Tropical Forest Habitat Loss. Sommer (1976) estimated the climatic range of the closed tropical moist forest biome to be 1,600 × 10⁶ hectares. Sommer's definition of moist forest corresponds to the more generic category of "closed tropical forest;" according to the Holdridge Lifezone Classification (Holdridge 1967) this includes closed dry, moist, wet, and rain forests. Some 97.7 percent of this forest is broadleaved, and in the following calculations only the broadleaved component is considered. Because of soil and climatic factors, a significant percentage of the climatic range of closed broadleaf tropical forest may have been historically covered with other vegetation (Myers 1980). If it is assumed that 10 percent of the potential closed forest zone was not forested (following Simberloff 1986a), then the extent of the biome prior to significant human impacts was 1,406 × 10⁶ hectares. In 1980, the total area of closed tropical forest was 1,081 × 10⁶ hectares; of this, 14.6 percent had been logged, leaving 923 × 10⁶ hectares of unlogged forest. (See Table 18.)

Natural forests may be disturbed in several different fashions. "Deforestation" refers to the transformation of forested land to permanently cleared land or to a shifting-cultivation cycle. Under shifting cultivation, some portion of this

Table 18. Area of Moist Tropical Closed Forest Biome (Thousand Hectares)

Region	Climatic Climax Area (Conifer and Broad-Leaf) (Sommer 1976)	Current Percent Coniferous (Lanly 1982)	Closed Broad-leaf Tropical Moist Forest				
			Climatic Climax Area[a]	Estimated Historical Area[b]	Total 1980 Area[c]	Undisturbed 1980 Area[d]	Legally Protected Forest (Lanly 1982)
Africa and Madagascar	362,000	3.6	348,968	314,071	204,622	164,008	9,018
Asia and Pacific	435,000	0.5	432,825	389,542	263,647	189,635	16,460
Latin America and Caribbean	803,000	2.8	780,516	702,464	613,110	569,403	13,906
Total	1,600,000	2.3	1,562,309	1,406,078	1,081,379	923,046	39,384

a. Calculated by assuming the current fraction of forests that are coniferous equals the historical fraction.
b. Calculated by subtracting an arbitrary 10 percent from the climatic broad-leaf climax area (after Simberloff 1986a).
c. Grainger 1984, based primarily on Lanly 1981.
d. Total area in 1980 minus categories "logged" and "managed" in Grainger 1984 (note that column headings in Grainger 1984, Table 2, are reversed).

"deforested" land will actually be under cover of woody vegetation as part of the fallow cycle. Forest may also be disturbed by temporary clearing during logging. Finally, forest that has already been disturbed by logging may be deforested if it is later transformed to shifting cultivation or permanently cleared.

It is difficult to predict deforestation rates in the coming decades. There is good reason to believe that they will increase because of population growth and, particularly in Latin America, because of increased access to forest resources. Satellite data for the southern portion of the Amazon basin of Brazil, for example, indicates that deforestation rates rose exponentially between 1975 and 1985 (Malingreau and Tucker 1988). Eventually, however, rates of forest loss will slow when the most accessible land has been cleared. Deforestation rates in Costa Rica, for instance, decreased between 1977 and 1983 as a result of the near complete deforestation of the country outside of protected areas (Sader and Joyce 1988). Lanly's (1982) projections of deforestation rates will be updated by FAO in 1990 based upon new surveys and analyses. Pending this new data, we estimate species extinction rates for

two deforestation rates: i) based on Lanly's (1982) projection for the period 1981 to 1985; and, ii) based on deforestation rates twice the projected rate of Lanly (1982). Actual rates of forest loss are likely to fall within these boundaries. The projected rate of deforestation or logging of undisturbed broadleaf forest between 1981 and 1985 is 7.4×10^6 hectares per year (Lanly 1982). *(See Table 19.)*

Species-Area Curve. The estimation of extinction rates is sensitive to the form of the species-area curve. Species-area curves, like that shown in Figure 8, generally fit closely to equations in the form:

$$S = cA^z,$$

where S = number of species, A = area, and c and z are constants. The exponent z determines the slope of the curve and is the critical parameter in estimating extinction rates. The slopes of species-area curves for various regions of tropical forests may differ because of differences in the numbers of habitats or life-zones present in the region. Moreover, slopes are likely to differ among various taxa due to differences in the average size of species' ranges. Species-area curves for groups that tend to have small

Table 19. Rate of Disturbance of Closed Broad-leaf Tropical Forest Per Year (Thousand Hectares)

	Undisturbed Forest		
	Area Logged[a]	Area Deforested[b]	Total Disturbed
Africa and Madagascar	635	290	925
Asia and Pacific	1,741	524	2,265
Latin America and Caribbean	1,960	2,281	4,241
Total	4,336	3,095	7,431

a. 1981–1985 projected logging rate of undisturbed closed broad-leaf tropical forest.

b. Sum of deforestation rates in categories: undisturbed productive forest and unproductive forest.

Source: Based on Lanly (1982).

ranges should have steeper slopes. Because of these difficulties, we model species extinction rates over a fairly broad range of slopes that are typically found in empirical studies of the species-area relationship ($0.15 < z < 0.35$; Connor and McCoy 1979).

Interpretation. If current trends continue, species loss over the next 30 years could amount to 3 to 17 percent of species currently present in closed tropical forests in Asia, Africa, and South America. *(See Table 5.)* If forest loss proceeds at twice the estimated rate, 6 to 44 percent of species could be lost in various regions. Table 5 includes estimates of the proportion of species that have *already* become extinct based on forest loss to date, but it must be remembered that these values are dependent upon an arbitrary estimate of the original extent of closed tropical forest and thus are imprecise.

Like any model, the species-area model is an approximation of reality that is based upon several assumptions. First, the model assumes that deforestation and logging eliminate all species originally present in the forest. In fact, many species may persist in a managed forest if logging intensity is not too high. Thus, the projected extinction rates could be reduced through forestry practices that ensure the maintenance of species populations in managed forests.

Second, the model assumes that extinction rates are unaffected by forest fragmentation. In practice, deforestation will convert many relatively continuous tracts of forest to a fragmented array of smaller patches and many of the species present in the patches may be lost if their populations are reduced below their minimum viable population size (Shaffer 1981). Exactly how fragmentation affects species richness within a region depends upon which areas remain under forest cover. Through the careful choice of areas to remain under natural forest cover, the potential loss of species within a region can be significantly reduced (Simberloff and Abele 1982, Quinn and Harrison 1988).

Finally, the model assumes that habitat loss occurs randomly among regions with various levels of species richness. Species loss could be reduced if species-rich regions are protected from deforestation.

Appendix 3: Protected Area Objectives

(Primary Objectives: +; Compatible Objectives: o)

Conservation Objective	Protected Area Designation (IUCN Category Number)										
	Scientific Reserve (1)	National Park (2)	Natural Monument (3)	Wildlife Reserve (4)	Protected Landscape (5)	Resource Reserve (6)	Anthropological Reserve (7)	Multiple-Use Area (8)	Extractive Reserve	Game Ranch	Recreation Area
Maintain sample ecosystems in natural state	+	+	+	+	o	o	+				o
Maintain ecological diversity and environment regulation	o	+	+	o	o	o	+	o	o	o	o
Conserve genetic resources	+	+	+	+	o	o	+	o	o	o	o
Provide education, research and environmental monitoring	+	o	+	+	o	o	o	o	o	o	o
Conserve watershed, flood control	o	+	o	o	o	o	o	o	o	o	o
Control erosion and sedimentation	o	o	o	o	o	o	o	o	o	o	o
Maintain indigenous uses or habitation					+	o	+	o	+		
Produce protein from wildlife				o	o	o	o	+	+	+	
Produce timber, forage or extractive commodities					o	o	o	+	+		
Provide recreation and tourism services		+	+	o	+		o	+	o		+
Protect sites and objects of cultural, historical, or archeological heritage		o	o	o	+	o	+	o	o	o	o
Protect scenic beauty	o	+	o	o	+			o	o	o	o
Maintain open options, management flexibility, multiple-use					o	+		+			
Contribute to rural development	o	+	o	o	+	o	o	+	+	+	o

Sources: Miller 1980, IUCN/UNEP 1986d.

Appendix 4: Glossary

Accession A sample of a crop variety collected at a specific location and time; may be of any size.

Adaptation A genetically determined characteristic that enhances an organism's ability to cope with its environment.

Allele One of several forms of the same gene.

Arthropods The animal phylum comprised of crustaceans, spiders, mites, centipedes, insects, and related forms. The largest of the phyla, containing more than three times the number of all other animal phyla combined.

Assemblage See "Community."

Avifauna All of the birds found in a given area.

Biodiversity The variety and variability of living organisms and the ecological complexes in which they occur; the variety of the world's species, including their genetic diversity and the assemblages they form.

Biogeochemical Cycles The movement of massive amounts of carbon, nitrogen, oxygen, hydrogen, calcium, sodium, sulfur, phosphorus, and other elements among various living and non-living components of the environment—including the atmosphere, soils, aquatic systems, and biotic systems—through the processes of production and decomposition.

Biogeography The scientific study of the geographic distribution of organisms.

Biome A major portion of the living environment of a particular region, such as a fir forest or grassland, characterized by its distinctive vegetation and maintained by local conditions of climate.

Biota All of the organisms, including animals, plants, fungi, and microorganisms, found in a given area.

Biotic Pertaining to any aspect of life, especially to characteristics of entire populations or ecosystems.

Center of Diversity Geographic region with high levels of genetic or species diversity.

Center of Endemism Geographic region with numerous locally endemic species.

Characteristic Diversity The pattern of distribution and abundance of populations, species, and habitats under conditions where humanity's influence on the ecosystem is no greater than that of any other biotic factor.

Class In taxonomy, a category just beneath the phylum and above the order; a group of related, similar orders.

Climax community End of a successional sequence; a community that has reached stability under a particular set of environmental conditions.

Cline Change in population characteristics over a geographical area, usually related to a corresponding environmental change.

Clone A population of individuals all derived asexually from the same single parent.

Community An integrated group of species inhabiting a given area. The organisms within a community influence one another's distribution, abundance, and evolution.

Conservation The management of human use of the biosphere so that it may yield the greatest sustainable benefit to present generations while maintaining its potential to meet the needs and aspirations of future generations.

Conservation of Biodiversity The management of human interactions with the variety of life forms and the complexes in which they occur so as to provide the maximum benefit to the present generation while maintaining their potential to meet the needs and aspirations of future generations.

Cosmopolitan Widely distributed over the globe.

Cryogenics The branch of physics relating to the effects and production of very low temperatures; as applied to living organisms, preservation in a dormant state by freezing, drying, or both.

Cultivar A cultivated variety (genetic strain) of a domesticated crop plant.

Decomposition The breakdown of organic materials by organisms in the environment, releasing energy and simple organic and inorganic compounds. About 10 percent of the energy that enters living systems through photosynthesis in plants passes to herbivores, and a fraction of this energy then passes to carnivores. Whether feeding on living or non-living material, however, the *detritivores* (the organisms consuming non-living material, such as many fungi, bacteria, and earthworms) and consumers break down organic material (such as sugars and proteins) to obtain energy for their own growth, thereby returning the inorganic components (the nutrients) to the environment, where they are again available to plants.

Demography The rate of growth and the age structure of populations, and the processes that determine these properties.

Donor Control A predator-prey interaction in which the predator does not control the prey population size.

Ecosystem The organisms of a particular habitat, such as a pond or forest, together with the physical environment in which they live; a dynamic complex of plant and animal communities and their associated non-living environment. Ecosystems have no fixed boundaries; instead, their parameters are set according to the scientific, management, or policy question being examined. Depending upon the purpose of analysis, a single lake, a watershed, or an entire region could be an ecosystem.

Ecotype A genetically differentiated subpopulation that is restricted to a specific habitat.

Endemic Restricted to a specified region or locality.

Environmental Heterogeneity The physical or temporal patchiness of the environment. Heterogeneity exists at all scales within natural communities, ranging from habitat differences between the top and underside of a leaf, to habitat patches created by treefalls within a forest, to the pattern of forests and grasslands within a region (Diamond 1980, Sousa 1984). The mosaic of habitat patches within an ecosystem is created by such disturbances as fire and storms; differences in microclimate, soils, and history; and both deterministic and random population variation. Patches in earlier stages of succession provide unique structural habitats and contain different species than those in late-successional stages do.

Evolution Any gradual change. Organic evolution is any genetic change in organisms from generation to generation.

Ex situ **Conservation** A conservation method that entails the removal of germplasm resources (seed, pollen, sperm, individual organisms) from their original habitat or natural environment.

Fauna All of the animals found in a given area.

Flora All of the plants found in a given area.

Frugivore An animal that eats fruit.

Gene The functional unit of heredity. The part of the DNA molecule that encodes a single enzyme or structural protein unit.

Gene Bank A facility established for the *ex situ* conservation of individuals (seeds), tissues, or reproductive cells of plants or animals.

Genetic diversity Variation in the genetic composition of individuals within or among species. The heritable genetic variation within and among populations.

Genotype The set of genes possessed by an individual organism.

Germplasm The genetic material, especially its specific molecular and chemical constitution, that comprises the physical basis of the inherited qualities of an organism.

Guild A group of organisms that share a common food resource.

Habitat The environment in which an organism lives. Habitat can also refer to the organisms and physical environment in a particular place.

Hybridization Crossing of individuals from genetically different strains, populations, or species.

Inbreeding A mating system involving the mating or breeding of closely related individuals, the most extreme form of which is self-fertilization. It is used to "fix" economically useful genetic traits in genetically improved populations; however, it also can result in fixation of deleterious recessive alleles.

Inbreeding Depression A reduction in fitness or vigor as a result of fixation of deleterious, recessive alleles from consistent inbreeding in a normally outbreeding population.

In situ Conservation A conservation method that attempts to preserve the genetic integrity of gene resources by conserving them within the evolutionary dynamic ecosystems of their original habitat or natural environment.

In vitro Storage of plant or animal germplasm in tissue-culture form in glass containers.

Landraces A crop cultivar or animal breed that evolved with and has been genetically improved by traditional agriculturalists, but has not been influenced by modern breeding practices.

Life form Characteristic structure of a plant or animal.

Minimum Viable Population The smallest isolated population having an x percent chance of remaining extant for y years despite the foreseeable effects of demographic, environmental, and genetic stochasticity and natural catastrophes. The probability of persistence and the time of persistence—x, and y—are often taken to be 99 percent and 1000 years respectively.

Mutualism Relationship between two or more species that benefits all parties.

Mycorrhizal Fungi A fungus living in a mutualistic association with plants and facilitating nutrient and water uptake.

Nectarivore An animal that eats nectar.

Nitrogen Fixation A process whereby atmospheric nitrogen is converted to nitrogen compounds that plants can utilize directly by *nitrogen fixing bacteria* living in mutualistic associations with plants.

Orthodox Seed Seed that can be dried to moisture levels between 4 and 6 percent and kept at low temperatures.

Pathogen A disease-causing microorganism; a bacterium or virus.

Phenotype The morphological, physiological, biochemical, behavioral, and other properties of an organism that develop through the interaction of genes and environment. *(See genotype.)*

Phylogenetic Pertaining to the evolutionary history of a particular group of organisms.

Phylum In taxonomy, a high-level category just beneath the kingdom and above the class; a group of related, similar classes.

Population A group of individuals with common ancestry that are much more likely to mate with one another than with individuals from another such group.

Predator Control A predator-prey interaction in which the predator controls the prey population size; that is, in which the predator population is the limiting factor for the prey population size.

Primary Productivity The transformation of chemical or solar energy to biomass. Most primary production occurs through photosynthesis, whereby green plants convert solar energy, carbon dioxide, and water to glucose and eventually to plant tissue. In addition, some bacteria in the deep sea can convert chemical energy to biomass through chemosynthesis. Primary production refers to the *amount* of material produced. *Net primary production* is the measure of the actual accumulation of biomass after some of the products of photosynthesis (or chemosynthesis) are expended for the plant's own maintenance. Productivity, or the *rate* of production, is affected by various environmental factors, including the amount of solar radiation, the availability of water and mineral nutrients, and temperature.

Protected Areas Legally established areas, under either public or private ownership, where the habitat is managed to maintain a natural or semi-natural state.

Recalcitrant Seed Seed that does not survive drying and freezing.

Rehabilitation The recovery of ecosystem services in a degraded ecosystem or habitat.

Restoration The return of an ecosystem or habitat to its original community structure and natural complement of species.

Selection Natural selection is the differential contribution of offspring to the next generation by various genetic types belonging to the same populations. Artificial selection is the intentional manipulation by man of the fitness of individuals in a population to produce a desired evolutionary response.

Species A population or series of populations of organisms that are capable of interbreeding freely with each other but not with members of other species.

Species Diversity A function of the distribution and abundance of species. Approximately synonymous with species richness. In more technical literature, includes considerations of the evenness of species abundances. An ecosystem is said to be more diverse, according to the more technical definition, if species present have equal population sizes and less diverse if many species are rare and some are very common.

Species Richness The number of species within a region. Commonly used as a measure of species diversity, but technically only one aspect of diversity.

Stability A function of several characteristics of community or ecosystem dynamics, including the degree of population fluctuations, the community's resistance to disturbances, the speed of recovery from disturbances, and the persistence of the community's composition through time.

Subspecies A subdivision of a species; a population or series of populations occupying a discrete range and differing genetically from other geographical races of the same species.

Succession The more or less predictable changes in the composition of communities following a natural or human disturbance. For example, after a gap is made in a forest by logging, clearing, fire, or treefall, the initial (or ''pioneer'') species are often fast-growing and shade-intolerant. These species are eventually replaced by shade-tolerant species that can grow beneath the pioneers. If a community is not further disturbed, the outcome of the successional sequence may be a so-called *climax* community whose composition is unchanging. In practice, many communities are frequently disturbed and may never reach a climax composition.

Systematics The study of the historical evolutionary and genetic relationships among organisms and of their phenotypic similarities and differences.

Taxon (pl. taxa) The named taxonomic unit (e.g. *Homo sapiens*, Hominidae, or Mammalia) to which individuals, or sets of species, are assigned. *Higher taxa* are those above the species level.

Taxonomy The naming and assignment of organisms to taxa.

Trophic Level Position in the food chain, determined by the number of energy-transfer steps to that level.

Variety *See Cultivar.*

Vascular Plants Plants with a well-developed vascular system that transports water, minerals, sugars, and other nutrients throughout the plant body. Excludes the bryophytes: mosses, hornworts, and liverworts.

References

Abramovitz, Janet N. In press. Report on the survey of U.S.-based efforts to research and conserve biological diversity in developing countries. World Resources Institute, Washington, DC.

Altieri, Miguel A., and Laura C. Merrick. 1987. In situ conservation of crop genetic resources through maintenance of traditional farming systems. *Economic Botany* 41:86–96.

Altieri, Miguel A., and Laura C. Merrick. 1988. Agroecology and in situ conservation of native crop diversity in the third world. *In:* E.O. Wilson and Francis M. Peter (eds.), *Biodiversity*, National Academy Press, Washington, DC, pp. 361–369.

Altieri, Miguel A., M. Kat Anderson, and Laura C. Merrick. 1987. Peasant agriculture and the conservation of crop and wild plant resources. *Conservation Biology* 1:49–58.

Antonovics, Janis, A.D. Bradshaw, and R.G. Turner. 1971. Heavy metal tolerance in plants. *Advances in Ecological Research* 7:1–85.

Arora, R.K. 1985. Genetic resources of less known cultivated food plants. Monograph No. 9. National Bureau of Plant Genetic Resources Science. New Delhi, India.

Ashton, Peter S. 1988. Conservation of biological diversity in botanical gardens. *In:* E.O. Wilson and Francis M. Peter (eds.), *Biodiversity*. National Academy Press, Washington, DC, pp. 269–278.

Avise, John C. and William S. Nelson. 1989. Molecular genetic relationships of the extinct dusky seaside sparrow. *Science* 243:646–648.

Bailey, Robert G., and Howard C. Hogg. 1986. A world ecoregions map for resource reporting. *Environmental Conservation* 13:195–202.

Barel, C.D.N., R. Dorit, P.H. Greenwood, G. Fryer, N. Hughes, P.B.N. Jackson, H. Kawanabe, R.H. Lowe-McConnell, M. Nagoshi, A.J. Ribbink, E. Trewavas, F. Witte, and K. Yamaoka. 1985. Destruction of fisheries in Africa's lakes. *Nature* 315:19–20.

Beehler, Bruce. 1985. Conservation of New Guinea rainforest birds. *In:* A.W. Diamond and T.E. Lovejoy (eds.), *Conservation of Tropical Forest Birds*. Technical Publication No. 4. International Council for Bird Preservation, Cambridge, U.K., pp. 233–247.

Bellrose, Frank C. 1976. *Ducks, Geese and Swans of North America*. Stackpole Books, Harrisburg, PA.

Benson, Delwin E. 1986. Game farming survey 2: sources of income. *Farmers Weekly*, 11 April:10–11, South Africa.

Bettini, Sergio (ed.). 1978. *Handbook of Experimental Pharmacology, Vol. 48: Arthropod Venoms*. Springer-Verlag, New York, NY.

Beven, S., F. Connor, and K. Beven. 1984. Avian biogeography in the Amazon basin and the biological model of diversification. *Journal of Biogeography* 11:383–399.

Biswell, Harold H. 1974. Effects of fire on chaparral. *In:* T.T. Kozlowski and C.E. Ahlgren, (eds.), *Fire and Ecosystems.* Academic Press, New York, NY, pp. 321–364.

Booth, William. 1987. Combing the earth for cures to cancer, AIDS. *Science* 237:969–970.

Booth, William. 1989. New thinking on old growth. *Science* 244:141–143.

Bormann, F.H. 1982. The effects of air pollution on the New England landscape. *Ambio* 11:338–346.

Brenan, J.P.M. 1978. Some aspects of the phytogeography of tropical Africa. *Annals of the Missouri Botanical Garden* 65:437–478.

Briggs, John C. 1974. *Marine Zoogeography.* McGraw-Hill, New York, NY.

Brown, James H., and Arthur C. Gibson. 1983. *Biogeography.* C.V. Mosby Company, St. Louis, MO.

Brush, Stephen B. 1989. Rethinking crop genetic resource conservation. *Conservation Biology* 3:19–29.

Bryson, Reid A., and Thomas J. Murray. 1977. *Climates of Hunger.* University of Wisconsin Press, Madison, WI.

Buddemeier, R.W., and S.V. Smith. 1988. Coral reef growth in an era of rapidly rising sea level: predictions and suggestions for long-term research. *Coral Reefs* 7:51–56.

Burnham, Charles R. 1988. The restoration of the American chestnut. *American Scientist* 76:478–487.

Byrne, Gregory. 1988. Cyclosporin turns five. *Science* 242:198.

Cassels, Richard. 1984. The role of prehistoric man in the faunal extinctions of New Zealand and other Pacific islands. *In:* Paul S. Martin and Richard G. Klein (eds.), *Quaternary Extinctions: A Prehistoric Revolution.* University of Arizona Press, Tucson, AZ, pp. 741–767.

Cayford, J.H. and D.J. McRae. 1983. The ecological role of fire in jack pine forests. *In:* Ross W. Wein and David A. MacLean (eds.), *The Role of Fire in Northern Circumpolar Ecosystems.* Scientific Committee on Problems of the Environment (SCOPE) No. 18. John Wiley and Sons, NY, pp. 183–199.

CPC (Center for Plant Conservation). 1988. CPC endangerment survey. CPC, 9 December 1989, Jamaica Plain, MA.

Chaney, William R., and Malek Basbous. 1978. The cedars of Lebanon, witnesses of history. *Economic Botany* 32:118–123.

Chapin, Mac. 1986. The Panamanian iguana renaissance. *Grassroots Development* 10(2):2–7.

Childress, James J., Horst Felback and George N. Somero. 1987. Symbiosis in the Deep Sea. *Scientific American* 256(5):114–120.

Clawson, David L. 1985. Harvest security and intraspecific diversity in traditional tropical agriculture. *Economic Botany* 39:56–67.

Cock, James H. 1982. Cassava: a basic energy source in the tropics. *Science* 218:755–762.

Colinvaux, Paul. 1987. Amazon diversity in light of the paleoecological record. *Quaternary Science Reviews* 6:93–114.

Colinvaux, Paul A. 1989. The past and future Amazon. *Scientific American* 260(5):102–108.

Connell, Joseph H. 1978. Diversity in tropical rain forests and coral reefs. *Science* 199:1302–1310.

Connor, Edward F., and Earl D. McCoy. 1979. The statistics and biology of the species-area relationship. *American Naturalist* 13:791–833.

Conway, William G. 1980. An overview of captive propagation. *In:* Michael E. Soulé and Bruce A. Wilcox. 1980. *Conservation Biology: An Evolutionary-Ecological Perspective.* Sinauer Associates, Inc., Sunderland, MA, pp. 199–208.

Conway, William G. 1988. Can technology aid species preservation? *In:* E.O. Wilson and Francis M. Peter (eds.), *Biodiversity.* National Academy Press, Washington, DC, pp. 263–268.

Crosby, Alfred W. 1986. *Ecological Imperialism: The Biological Expansion of Europe, 900–1900.* Cambridge University Press, Cambridge, U.K.

D'Arcy, W.G. 1977. Endangered landscapes in Panama and Central America: the threat to plant species. *In:* Ghillean T. Prance and Thomas S. Elias (eds.), *Extinction is Forever.* New York Botanical Garden, Bronx, NY, pp. 89–104.

Daniel, J.G., and A. Kulasingam. 1974. Problems arising from large-scale forest clearing for agricultural use. *Malaysian Forester* 37(3):152–160.

Darus, Bahauddin. 1983. The management and development of southeast Asian small-scale fisheries and the example of the Bubun coastal village development project, North Sumatra Province, Indonesia. *In:* Kenneth Ruddle and R.E. Johannes (eds.), *The Traditional Knowledge and Management of Coastal Systems in Asia and the Pacific.* United Nations Educational, Scientific and Cultural Organization (UNESCO), Regional Office for Science and Technology for Southeast Asia, Jakarta Pusat, Indonesia, pp. 209–228.

Dasmann, R.F. 1973. A system for defining and classifying natural regions for purposes of conservation. Occasional Paper 7. International Union for Conservation of Nature and Natural Resources (IUCN), pp. 1–47.

Davis, Margaret B., and Catherine Zabinski. Changes in geographical range resulting from greenhouse warming: effects on biodiversity in forests. *In:* R.L. Peters and T.E. Lovejoy (eds.), *Consequences of the Greenhouse Effect for Biological Diversity.* Yale University Press, New Haven CT. In press.

Davis, Stephen D., Stephen J.M. Droop, Patrick Gregerson, Louise Henson, Christine J. Leon, Jane L. Villa-Lobos, Hugh Synge, and Jana Zantovska. 1986. *Plants in Danger: What Do We Know?* International Union for Conservation of Nature and Natural Resources, Gland, Switzerland.

Dawson, William R, J. David Ligon, Joseph R. Murphy, J.P. Myers, Daniel Simberloff, and Jared Verner. 1987. Report of the scientific advisory panel on the spotted owl. *Condor* 89:205–229.

Day, David. 1981. *The Doomsday Book of Animals.* Viking Press, New York, NY.

de Alba, Jorge. 1987. Criollo cattle of Latin America. *In:* John Hodges (ed.), *Animal Genetic Resources: Strategies for Improved Use and Conservation.* Animal Production and Health Paper 66. United Nations Food and Agriculture Organization, Rome, pp. 19–43.

DeBach, Paul. 1974. *Biological Control by Natural Enemies.* Cambridge University Press, New York, NY.

Delcourt, Hazel R. 1987. The impact of prehistoric agriculture and land occupation on natural vegetation. *Trends in Ecology and Evolution* 2(2):39–44.

Dewar, Robert E. 1984. Extinctions in Madagascar. *In:* Paul S. Martin and Richard G. Klein (eds.), *Quaternary Extinctions: A Prehistoric Revolution.* University of Arizona Press, Tucson, AZ, pp. 574–593.

Diamond, A.W. 1985. The selection of critical areas and current conservation efforts in tropical forest birds. *In:* A.W. Diamond and T.E. Lovejoy (eds.), *Conservation of Tropical Forest Birds.* Technical Publication No. 4. International Council for Bird Preservation, Cambridge, U.K., pp. 33–48.

Diamond, Jared M. 1980. Patchy distributions of tropical birds. *In:* Michael E. Soulé and Bruce A. Wilcox (eds.), *Conservation Biology: An Evolutionary-Ecological Perspective.* Sinauer Associates, Sunderland, MA, pp. 57–74.

Diamond, Jared M. 1984. ''Normal'' extinctions of isolated populations. *In:* Matthew H. Nitecki (ed.), *Extinctions.* University of Chicago Press, Chicago, IL, pp. 191–246.

Diamond, Jared. 1988a. Factors controlling species diversity: overview and synthesis. *Annals of the Missouri Botanical Gardens* 75:117–129.

Diamond, Jared M. 1988b. Red books or green lists? *Nature* 332:304–305.

Dobson, Andrew, and Robin Carper. In press. Global warming and potential changes in host-parasite and disease-vector relationships. *In:* R.L. Peters and T.E. Lovejoy (eds.), *Consequences of the Greenhouse Effect for Biological Diversity.* Yale University Press, New Haven, CT.

Dodson, Calaway H. 1975. Coevolution of orchids and bees. *In:* Lawrence E. Gilbert and Peter H. Raven (eds.), *Coevolution of Animals and Plants.* University of Texas Press, Austin, TX, pp. 91–99.

Dourojeanni, Marc J. 1985. Over-exploited and under-used animals in the Amazon region. *In:* Ghillean T. Prance and Thomas E. Lovejoy (eds.), *Amazonia.* Pergamon Press, Oxford, pp. 419–433.

Dresser, Betsy L. 1988. Cryobiology, embryo transfer, and artificial insemination in ex situ animal conservation programs. *In:* E.O.

Wilson and Francis M. Peter (eds.), *Biodiversity.* National Academy Press, Washington, DC, pp. 296–308.

ESCAP (Economic and Social Commission for Asia and the Pacific). 1985. *Marine Environmental Problems and Issues in the ESCAP Region.* United Nations, ESCAP, Bangkok, Thailand.

Ehrlich, Paul R., and Anne H. Ehrlich. 1981. *Extinction.* Random House, New York, NY.

Ehrlich, Paul R., and Harold A. Mooney. 1983. Extinction, substitution, and ecosystem services. *BioScience* 33:248–254.

Eisner, Thomas. Chemical exploration of nature: a proposal for action. *In: Ecology, Economics, and Ethics: The Broken Circle.* Yale University Press, New Haven, CT. In press.

Elton, Charles. 1930. *Animal Ecology and Evolution.* Clarendon Press, Oxford.

Endler, John A. 1982. Pleistocene forest refuges: fact or fancy? *In:* Ghillean T. Prance (ed.), *Biological Diversification in the Tropics.* Columbia University Press, New York, NY, pp. 641–657.

Erwin, Terry L. 1982. Tropical forests: their richness in Coleoptera and other Arthropod species. *Coleopterists Bulletin* 36:74–75.

Erwin, Terry L., and Janice C. Scott. 1980. Seasonal and size patterns, trophic structure and richness of Coleoptera in the tropical arboreal ecosystem: the fauna of the tree *Luehea seemannii* Triana and Planch in the Canal Zone of Panama. *Coleopterists Bulletin* 34(3):305–322.

Estes, James A., and John F. Palmisano. 1974. Sea otters: their role in structuring nearshore communities. *Science* 185:1058–1060.

Estes, James A., Norman S. Smith, and John F. Palmisano. 1978. Sea otter predation and community organization in the western Aleutian Islands, Alaska. *Ecology* 59:822–833.

FAO (Food and Agriculture Organization of the United Nations). 1981. Utilization of fish and its role in human nutrition. Committee on Fisheries, 14th Session, FAO, Rome.

FAO. 1982. Management and utilization of mangroves in Asia and the Pacific. Environment Paper 3. FAO, Rome.

FAO. 1984. Fish in nutrition. WFC/NF/84/2 FAO, Rome.

FAO. 1985a. Mangrove management in Thailand, Malaysia, and Indonesia. Environment Paper 4. FAO, Rome.

FAO. 1985b. Status of in situ conservation of plant genetic resources. Commission on Plant Genetic Resources, First Session, FAO, Rome.

FAO. 1986. Data book on endangered tree and shrub species and provenances. Forestry Paper 77. Forest Resources Division, FAO, Rome.

FAO. 1987a. *1986 FAO Production Yearbook.* Vol. 40. FAO, Rome.

FAO. 1987b. *World Food Report 1987.* FAO, Rome.

FAO. 1989. *Forest Products Yearbook 1976–1987.* FAO, Rome.

FAO/SIDA (FAO/Swedish International Development Authority). 1987. *Restoring the Balance: Women and Forest Resources.* FAO, Rome.

Farnsworth, Norman R. 1988. Screening plants for new medicines. *In:* E.O. Wilson and Francis M. Peter (eds.), *Biodiversity.* National Academy Press, Washington, DC, pp. 83–97.

Fischer, Alfred G. 1960. Latitudinal variations in organic diversity. *Evolution* 14:64–81.

Fonseca, Gustavo A.B. da. 1985. The vanishing Brazilian Atlantic forest. *Biological Conservation* 34:17–34.

Foose, Thomas J. 1983. The relevance of captive populations to the conservation of biodiversity. *In:* Christine M. Schonewald-Cox, Steven M. Chambers, Bruce MacBryde, and W. Lawrence Thomas (eds.), *Genetics and Conservation.* Benjamin/Cummings Publishing Co., Menlo Park, CA, pp. 374–401.

Foster, Robin B. 1980. Heterogeneity and disturbance in tropical vegetation. *In:* Michael E. Soulé and Bruce A. Wilcox (eds.), *Conservation Biology: An Evolutionary-Ecological Perspective.* Sinauer Associates, Sunderland, MA, pp. 75–92.

Frankel, O.H., and Michael E. Soulé. 1981. *Conservation and Evolution.* Cambridge University Press, Cambridge, U.K.

Franklin, Ian Robert. 1980. Evolutionary change in small populations. *In:* Michael E. Soulé and Bruce A. Wilcox. *Conservation Biology: An Evolutionary-Ecological Perspective.* Sinauer Associates, Sunderland, MA, pp. 135–149.

Franklin, Jerry F. 1988. Structural and functional diversity in temperate forests. *In:* E.O. Wilson and Francis M. Peter (eds.), *Biodiversity.* National Academy Press, Washington, DC, pp. 166–175.

GAO (General Accounting Office of the United States). 1988. *Endangered Species: Management Improvements Could Enhance Recovery Program.* GAO, Washington, DC.

Gentry, Alwyn H. 1982. Neotropical floristic diversity: phytogeographical connections between Central and South America, Pleistocene climatic fluctuations, or an accident of the Andean orogeny? *Annals of the Missouri Botanical Garden* 69:557–593.

Gentry, Alwyn H. 1986. Endemism in tropical versus temperate plant communities. *In:* Michael E. Soulé (ed.), *Conservation Biology: The Science of Scarcity and Diversity.* Sinauer Associates, Sunderland, MA, pp. 153–181.

Gentry, Alwyn H. 1988. Tree species richness of upper Amazonian forests. *Proceedings of the National Academy of Sciences* 85:156–159.

Gilbert, Lawrence E. 1980. Food web organization and the conservation of neotropical diversity. *In:* Michael E. Soulé and Bruce A. Wilcox (eds.), *Conservation Biology: An Evolutionary-Ecological Perspective.* Sinauer Associates, Sunderland, MA, pp. 11–33.

Goldman, Barry, and Frank Hamilton Talbot. 1976. Aspects of the ecology of coral reef fishes. *In:* O.A. Jones and R. Endean (eds.), *Biology and Geology of Coral Reefs Vol. 3.* Academic Press, New York, NY, pp. 125–154.

Goodman, Daniel. 1987. How do any species persist? Lessons for conservation biology. *Conservation Biology* 1:59–62.

Graham, Russell W. 1986. Plant-animal interactions and Pleistocene extinctions. *In:* D.K. Elliott (ed.), *Dynamics of Extinction.* John Wiley and Sons, New York, NY, pp. 131–154.

Graham, Russell W. 1988. The role of climatic change in the design of biological reserves: the paleoecological perspective for conservation biology. *Conservation Biology* 2:391–394.

Grainger, Alan. 1984. Quantifying changes in forest cover in the humid tropics: overcoming current limitations. *Journal of World Forest Resource Management* 1:3–63.

Grassle, J. Frederick. 1989. Species diversity in deep-sea communities. *Trends in Ecology and Evolution* 4(1):12–15.

Grassle, J. Frederick, Nancy J. Maciolek, and James A. Blake. Are deep sea communities resilient? *In:* George M. Woodwell (ed.), *The Earth in Transition: Patterns and Processes of Biotic Impoverishment.* Proceedings of a Conference held at Woods Hole Research Center, October 1986, Woods Hole, MA. In press.

Gregg, William P., Jr., and Betsy Ann McGean. 1985. Biosphere reserves: their history and their promise. *Orion* 4(3):40–51.

Grigg, Richard W., and David Epp. 1989. Critical depth for the survival of coral islands: effects on the Hawaiian archipelago. *Science* 243:638–641.

Gulick, P., C. Hershey, and J. Esquinas Alcazar. 1983. *Genetic Resources of Cassava and Wild Relatives.* International Board for Plant Genetic Resources 82/111, Rome.

Haffer, Jürgen. 1969. Speciation in Amazon forest birds. *Science* 165:131–137.

Hairston, Nelson G., F.E. Smith, and L.B. Slobodkin. 1960. Community structure, population control and competition. *American Naturalist* 94:421–425.

Hansen, Michael. 1987. *Alternatives to Pesticides in Developing Countries.* Institute for Consumer Policy Research, Mount Vernon, NY.

Haq, Nazmul. 1988. Planting the future. *International Agricultural Development* September/October:13–15.

Harlan, J.R. 1975. *Crops and Man.* American Society of Agronomy, Inc. and the Crop Science Society of America, Inc., Madison, WI.

Hargrove, Thomas R., Victoria L. Cabanilla, and W. Ronnie Coffman. 1988. Twenty years of rice breeding. *BioScience* 38:675–681.

Hartshorn, Gary S., Robert Simeone, and Joseph A. Tosi, Jr. 1987. Sustained yield management of tropical forests: a synopsis of the Palcazu development project in the Central Selva of the Peruvian Amazon. *In:* J.C. Figueroa C., F.H. Wadsworth and S. Branham (eds.), *Management of the Forests of Tropical America: Prospects and Technologies.* Institute of Tropical Forestry, Rio Piedras, Puerto Rico, pp. 235–243.

Hartshorn, Gary S. 1989. Application of gap theory to tropical forest management: natural regeneration on strip clear-cuts in the Peruvian Amazon. *Ecology* 70:567–569.

Hawkes, J.G. 1983. *The Diversity of Crop Plants.* Harvard University Press, Cambridge, MA.

Hayden, Bruce P., G. Carleton Ray, and Robert Dolan. 1984. Classification of coastal and marine environments. *Environmental Conservation* 11:199–207.

Heywood, V[ernon] H. 1987. The changing role of the botanic garden. *In:* D. Bramwell, O. Hamann, V. Heywood, and H. Synge (eds.), *Botanic Gardens and the World Conservation Strategy.* Academic Press, London, U.K., pp. 3–18.

Hillman-Smith, Kes, Mankoto ma Oyisenzoo, and Fraser Smith. 1986. A last chance to save the northern white rhino? *Oryx* 20:20–26.

Hoeck, H.N. 1982. Population dynamics, dispersal and genetic isolation in two species of Hyrax (*Heterohyrax brucei* and *Procavia johnstoni*) on habitat islands in the Serengeti. *Zeitschrift für Tierpsychologie* 59:177–210.

Hoeck, H.N. 1984. Introduced fauna. *In:* R. Perry (ed.), *Key Environments: Galapagos.* Pergamon Press, Oxford, U.K., pp. 233–245.

Holdridge, L[eslie] R. 1967. *Life Zone Ecology.* Tropical Science Center, San Jose, Costa Rica.

Hourigan, Thomas F., and Reese, Ernst S. 1987. Mid-ocean isolation and the evolution of Hawaiian reef fishes. *Trends in Ecology and Evolution* 2(7):187–191.

Huber, Otto, and Clara Alarcorn. 1988. *Mapa de Vegetación de Venezuela.* Ministerio del Ambiente y de los Recursos Naturales Renovables, Caracas, Venezuela.

Hubbell, Stephen P. 1979. Tree dispersion, abundance, and diversity in a tropical dry forest. *Science* 203:1299–1309.

Hubbell, Stephen P., and Robin B. Foster. 1985. Biology, chance, and history, and the structure of tropical forest tree communities. *In:* Jared Diamond and Ted J. Case (eds.), Community Ecology. Harper & Row, New York, NY, pp. 314–329.

Hubbell, Stephen P., and Robin B. Foster. 1986. Commonness and rarity in a neotropical forest: implications for tropical tree conservation. *In:* Michael E. Soulé (ed.), *Conservation Biology: The Science of Scarcity and Diversity.* Sinauer Associates, Sunderland, MA, pp. 205–231.

Hunter, Malcolm L. Jr., George L. Jacobson Jr., and Thompson Webb III. 1988. Paleoecology and the coarse-filter approach to maintaining biological diversity. *Conservation Biology* 2:375–385.

IUCN (International Union for Conservation of Nature and Natural Resources). 1980. *World Conservation Strategy.* IUCN/UNEP/WWF, Gland, Switzerland.

IUCN. 1983. *The IUCN Invertebrate Red Data Book.* IUCN, Gland, Switzerland.

IUCN. 1984. Categories, objectives and criteria for protected areas. *In:* Jeffrey A. McNeely and Kenton R. Miller (eds.), *National Parks, Conservation and Development.* Smithsonian Institution Press, Washington, DC, pp. 47–53.

IUCN. 1985. *United Nations List of National Parks and Protected Areas.* IUCN, Gland, Switzerland, and Cambridge, U.K.

IUCN. 1987. Elephant population estimates, trends, ivory quotas and harvests. African Elephant and Rhino Specialist Group, Report to the Sixth Meeting of the Conference of the Parties (CITES), July 1987, Ottawa, Canada. IUCN, Gland, Switzerland.

IUCN. 1988. *1988 IUCN Red List of Threatened Animals.* IUCN, Gland, Switzerland.

IUCN/UNEP (IUCN/United Nations Environment Programme). 1986a. *Review of the Protected Areas System in the Indo-Malayan Realm.* IUCN, Gland, Switzerland.

IUCN/UNEP. 1986b. *Review of the Protected Areas System in the Afrotropical Realm.* IUCN, Gland, Switzerland.

IUCN/UNEP. 1986c. *Review of the Protected Areas System in Oceania.* IUCN, Gland, Switzerland.

IUCN/UNEP. 1986d. *Managing Protected Areas in the Tropics.* IUCN, Gland, Switzerland, and Cambridge, U.K.

IUCN/UNEP. 1988. *Coral Reefs of the World.* 3 Volumes. IUCN, Gland, Switzerland and Cambridge, U.K./UNEP, Nairobi, Kenya.

IUCN/WWF (IUCN/Worldwide Fund for Nature). 1989. *The Botanic Gardens Conservation Strategy.* IUCN, Gland, Switzerland.

Isaac, Erich. 1970. *Geography of Domestication.* Prentice-Hall, Inc., Englewood Cliffs, NJ.

Jacobson, George L. Jr., Thompson Webb III, Eric C. Grimm. 1987. Patterns and rates of vegetation change during the deglaciation of eastern North America. *The Geology of North America: North America and Adjacent Oceans During the Last Deglaciation* Vol. K–3, Geological Society of America, Boulder, CO.

James, D. 1984. The future for fish in nutrition. *Infofish Marketing Digest* 4(84):41–44.

Janzen, Daniel H. 1979. How to be a fig. *Annual Review of Ecology and Systematics* 10:13–51.

Janzen, Daniel H. 1986. *Guanacaste National Park: Tropical Ecological and Cultural Restoration.* Editorial Universidad Estatal a Distancia, San José, Costa Rica.

Janzen, Daniel H. 1988a. Tropical dry forests: the most endangered major tropical ecosystem. *In:* E.O. Wilson and Francis M. Peter (eds.), *Biodiversity.* National Academy Press, Washington, DC, pp. 130–137.

Janzen, Daniel H. 1988b. Management of habitat fragments in a tropical dry forest: growth. *Annals of the Missouri Botanical Garden* 75:105–116.

Janzen, Daniel H., W. Hallwachs, W.A. Haber, I. Chacon, A. Chacon, M.M. Chaverria, M. Espinosa, E. Araya, and F.G. Stiles. Toward a definitive checklist of the Costa Rican moths in the family Sphingidae. Unpublished manuscript.

Jenkins, M.D. (ed.). 1987. *Madagascar: An Environmental Profile.* International Union for the Conservation of Nature and Natural Resources, Gland, Switzerland.

Johannes, R.E., and B.G. Hatcher. 1986. Shallow tropical marine environments. *In:* Michael E. Soulé (ed.), *Conservation Biology: The Science of Scarcity and Diversity.* Sinauer Associates Inc., Sunderland, MA, pp. 371–382.

Johns, Andrew D. 1985. Selective logging and wildlife conservation in tropical rain-forest: problems and recommendations. *Biological Conservation* 31:355–375.

Johns, Andrew D. 1986. *Effects of Habitat Disturbance on Rainforest Wildlife in Brazilian Amazon.* Final report, World Wildlife Fund U.S. (WWF) project US–302. WWF, Washington, DC.

Johns, Andrew D. 1988. Economic development and wildlife conservation in Brazilian Amazonia. *Ambio* 17:302–306.

Jokiel, Paul L. 1987. Ecology, biogeography, and evolution of corals in Hawaii. *Trends in Ecology and Evolution* 2(7):179–182.

Jordan, William R. III, Robert L. Peters II, and Edith B. Allen. 1988. Ecological restoration as a strategy for conserving biological diversity. *Environmental Management* 12:55–72.

Juma, Calestous. 1989. *The Gene Hunters: Biotechnology and the Scramble for Seeds.* Princeton University Press, Princeton, NJ.

Karr, J.R. 1982. Avian extinction on Barro Colorado Island, Panama: a reassessment. *American Naturalist* 119:220–239.

Kloppenburg, Jack, Jr., and Daniel Lee Kleinman. 1987. The plant germplasm controversy. *BioScience* 37(3):190–198.

Kloppenburg, Jack Ralph Jr. 1988a. *First the Seed.* Cambridge University Press, Cambridge, MA.

Kloppenburg, Jack R. Jr. 1988b. *Seeds and Sovereignty: Debate Over the Use and Control of Plant Genetic Resources.* Duke University Press, Durham, NC.

Knoll, Andrew H. 1986. Patterns of change in plant communities through geological time. *In:* Jared Diamond and Ted J. Case (eds.), *Community Ecology.* Harper and Row, New York, NY, pp. 126–141.

Lande, Russell. 1988. Genetics and demography in biological conservation. *Science* 241:1455–1460.

Lanly, Jean-Paul. 1981. *Tropical Forest Resources Assessment Project (GEMS): Tropical Africa, Tropical Asia, Tropical America.* (4 vols.). FAO/UNEP, Rome.

Lanly, Jean-Paul. 1982. Tropical forest resources. Forestry Paper No. 30, Food and Agriculture Organization of the United Nations, Rome.

Lawton J.H., and S.L. Pimm. 1978. Population dynamics and the length of food chains. *Nature* 272:190.

Leigh, J.H., J.D. Briggs, and W. Hartley. 1982. The conservation status of Australian plants. *In:* R.H. Groves, and W.D.L. Ride (eds.), *Species at Risk: Research in Australia,* Springer-Verlag, New York, NY, pp. 13–25.

Lewis, D.M., G.B. Kaweche, and A. Mwenya. 1987. Wildlife conservation outside protected areas: lessons from an experiment in Zambia. Publication No. 4. Lupande Research Project, National Parks and Wildlife Service, Lusaka, Zambia.

Lovejoy, Thomas E. 1980. A projection of species extinctions. *In:* Council on Environmental Quality (CEQ), *The Global 2000 Report to the President, Vol. 2.* CEQ, Washington, DC., pp. 328–331.

Lovejoy, T[homas] E., R.O. Bierregaard Jr., A.B. Rylands, J.R. Malcolm, C.E. Quintela, L.H. Harper, K.S. Brown Jr., A.H. Powell, G.V.N. Powell, H.O.R. Schubart, and M.B. Hays. 1986. Edge and other effects of isolation on Amazon forest fragments. *In:* Michael E. Soulé (ed.), *Conservation Biology: The Science of Scarcity and Diversity.* Sinauer Associates, Sunderland, MA, pp. 257–285.

Lucas, Gren, and Hugh Synge. 1978. *The IUCN Plant Red Data Book.* International Union for the Conservation of Nature and Natural Resources, Gland, Switzerland.

Lugo, Ariel E., and Samuel C. Snedaker. 1974. The ecology of mangroves. *Annual Review of Ecology and Systematics* 5:39–64.

Lyman, Judith M. 1984. Progress and planning for germplasm conservation of major food crops. *Plant Genetic Resources Newsletter* 60:3–21.

MacKenzie, James J., and Mohamed T. El-Ashry. 1988. *Ill Winds: Airborne Pollution's Toll on Trees and Crops.* World Resources Institute, Washington, DC.

Machlis, Gary E., and David L. Tichnell. 1985. *The State of the World's Parks.* Westview Press, Boulder, CO.

Malingreau, Jean-Paul, and Compton J. Tucker. 1988. Large-scale deforestation in the southeastern Amazon basin of Brazil. *Ambio* 17:49–55.

Maltby, Edward. 1988. Wetland resources and future prospects: an international perspective. *In:* John Zelazny and J. Scott Feierabend (eds.), *Wetlands: Increasing Our Wetland Resources.* National Wildlife Federation, Washington, DC, pp. 3–14.

Martin, Paul S. 1973. The discovery of America. *Science* 179:969–974.

Martin, Paul S. 1986. Refuting late pleistocene extinction models. *In:* D.K. Elliott (ed.), *Dynamics of Extinction.* John Wiley and Sons, New York, NY, pp. 107–130.

Martin, Paul S., and Richard G. Klein (eds.). 1984. *Quaternary Extinctions: a Prehistoric Revolution.* University of Arizona Press. Tucson, AZ.

May, Robert M. 1981. Patterns in multi-species communities. *In:* Robert M. May (ed.), *Theoretical Ecology: Principles and Applications* 2nd Edition. Blackwell Scientific Publications, Oxford, U.K., pp. 197–227.

May, Robert M. 1985. Evolution of pesticide resistance. *Nature* 315:12–13.

May, Robert M. 1988. How many species are there on earth? *Science* 241:1441–1449.

McCormick, J. Frank, and Robert B. Platt. 1980. Recovery of an Appalachian forest following the chestnut blight, or, Catherine Keever—you were right! *American Midland Naturalist* 104:264–273.

McManus, John W. 1988. Coral reefs of the ASEAN region: status and management. *Ambio* 17:189–193.

McNaughton, S.J. 1985. Ecology of a grazing ecosystem: the Serengeti. *Ecological Monographs* 55:259–294.

McNeely, Jeffrey A. 1988. *Economics and Biological Diversity.* International Union for the Conservation of Nature and Natural Resources, Gland, Switzerland.

McNeely, Jeffrey A. 1989. Common property resource management or government ownership: improving the conservation of biological resources. Paper presented at Conference on Incentives and Constraints: Macroeconomic Policy Impacts on Natural Resource Utilization. Smithsonian Institution, 12 May 1989, Washington, DC.

McNeely, Jeffrey A., Kenton R. Miller, Walter V. Reid, Russell A. Mittermeier and Timothy B. Werner. 1989. *Conserving the World's Biological Diversity.* World Resources Institute/International Union for Conservation of Nature and Natural Resources/Conservation International/World Bank. Washington, DC, and Gland, Switzerland.

Meith, Nikki, and Richard Helmer. 1983. Marine environment and coastal resources in southeast Asia: a threatened heritage. *In:* Elisabeth Mann Borgese and Norton Ginsburg (eds.), *Ocean Yearbook 4.* University of Chicago Press, Chicago, IL, pp. 260–294.

Melillo, J.M., C.A. Palm, R.A. Houghton, and G.M. Woodwell. 1985. A comparison of two recent estimates of disturbance in tropical forests. *Environmental Conservation* 12:37–40.

Miller, Constance I. 1987. The California forest germplasm conservation project: a case for genetic conservation of temperate tree species. *Conservation Biology* 1:191–193.

Miller, Kenton R. 1975. Guidelines for the management and development of national parks and reserves in the American humid tropics. *In:* Proceedings of a Meeting on the Use of Ecological Guidelines for Development in the American Humid Tropics,

Caracas 20–22 February, 1974. International Union for Conservation of Nature and Natural Resources, Morges, Switzerland, pp. 94–105.

Miller, Kenton R. 1980. Planificación de parques nacionales para el ecodesarrollo en Latinoamérica. Fundación para la Ecología y para la Protección del Medio Ambiente, Madrid.

Miller, Kenton, Walter Reid, and Jeffrey McNeely. 1989. A global strategy for conserving biodiversity. *Diversity* 5(1):4–7.

Miller, R. Michael. 1987. Mycorrhizae and succession. *In:* William R. Jordan III, Michael E. Gilpin, and John D. Aber (eds.), *Restoration Ecology: A Synthetic Approach to Ecological Research.* Cambridge University Press, Cambridge, U.K., pp. 205–219.

Mooney, H[arold] A., S.P. Hamburg, and J.A. Drake. 1986. The invasions of plants and animals into California. *In:* H.A. Mooney and J.A. Drake (eds.), *Ecology of Biological Invasions of North America and Hawaii.* Ecological Studies #58. Springer-Verlag, New York, NY, pp. 250–272.

Mooney, Harold A. 1988. Lessons from Mediterranean-climate regions. *In:* E.O. Wilson and Francis M. Peter (eds.), *Biodiversity.* National Academy Press, Washington, DC, pp. 157–165.

Mohsin, A.K. Mohammad, and Mohd Azmi Ambak. 1983. *Freshwater Fishes of Peninsular Malaysia.* University Pertanian Malaysia Press, Kuala Lumpur, Malaysia.

Mori, Scott A., Brian M. Boom, André M. de Carvalho and Talmón S. dos Santos. 1983. Southern Bahian moist forests. *Botanical Review* 49:155–232.

Murphy, Peter G., and Ariel E. Lugo. 1986. Ecology of tropical dry forest. *Annual Review of Ecology and Systematics* 17:67–88.

Myers, K. 1986. Introduced vertebrates in Australia, with emphasis on the mammals. *In:* R.H. Groves and J.J. Burdon (eds.), *Ecology of Biological Invasions.* Cambridge University Press, Cambridge, U.K., pp. 120–136.

Myers, Norman. 1979. *The Sinking Ark: A New Look at the Problem of Disappearing Species.* Pergamon Press, Oxford, U.K.

Myers, Norman. 1980. *Conversion of Tropical Moist Forests.* The National Research Council, National Academy of Sciences, Washington, DC.

Myers, Norman. 1983. *A Wealth of Wild Species.* Westview Press, Boulder, CO.

Myers, Norman. 1988. Threatened biotas: ''hotspots'' in tropical forests. *The Environmentalist* 8(3):1–20.

Naveh, Zev, and R.H. Whittaker. 1979. Structural and floristic diversity of shrublands and woodlands in northern Israel and other Mediterranean areas. *Vegetatio* 41:171–190.

Nilsson, Greta. 1983. *The Endangered Species Handbook.* Animal Welfare Institute, Washington, DC.

Norton, Bryan G. 1987. *Why Preserve Natural Variety?* Princeton University Press, Princeton, NJ.

Noss, Reed F. 1983. A regional landscape approach to maintain diversity. *BioScience* 33:700–706.

NRC (National Research Council). 1975. *Underexploited Tropical Plants with Promising Economic Value.* National Academy of Sciences, Washington, DC.

NRC. 1977. *Leucaena: Promising Forage and Tree Crop for the Tropics.* National Academy of Sciences, Washington, DC.

NRC. 1979. *Tropical Legumes: Resources for the Future.* National Academy of Sciences, Washington, DC.

NRC. 1980. *Research Priorities in Tropical Biology.* National Academy of Sciences, Washington, DC.

NRC. 1982. *Ecological Aspects of Development in the Humid Tropics.* National Academy Press, Washington, DC.

NRC. 1983. *Little-Known Asian Animals with a Promising Economic Future.* National Academy Press, Washington, DC.

NRC. 1986. *Ecological Knowledge and Environmental Problem Solving.* National Academy Press, Washington, DC.

OAS (Organization of American States), Government of Peru, and United Nations Environment Programme. 1987. *Minimum Conflict: Guidelines for Planning the Use of American Humid Tropic Environments.* OAS, Washington, DC.

OAS. 1988. *Inventory of Caribbean Marine and Coastal Protected Areas.* OAS, Washington, DC.

O'Brien, S.J., M.E. Roelke, L. Marker, A. Newman, C.A. Winkler, D. Meltzer, L. Colly, J.F. Evermann, M. Bush, and D.E. Wildt. 1985. Genetic basis for species vulnerability in the cheetah. *Science* 227:1428–1434.

Ocana, Gilberto, Ira Rubinoff, Nicholas Smythe, and Dagmar Werner. 1988. Alternatives to destruction: research in Panama. *In:* E.O. Wilson and Francis M. Peter (eds.), *Biodiversity.* National Academy Press, Washington, DC, pp. 370–376.

Odum, William E. 1982. Environmental degradation and the tyranny of small decisions. *BioScience* 32:728–729.

Oksanen, Lauri. 1988. Ecosystem organization: mutualism and cybernetics or plain Darwinian struggle for existence? *American Naturalist* 131:424–444.

Oldfield, Margery L. 1984. *The Value of Conserving Genetic Resources.* U.S. Department of Interior, National Park Service, Washington, DC.

Oldfield, Margery L., and Janis B. Alcorn. 1987. Conservation of traditional agroecosystems. *BioScience* 37:199–208.

Olson, Storrs L. 1989. Extinction on islands: man as a catastrophe. *In:* M. Pearl, and D. Western (eds.). *Conservation Biology for the Next Century.* Oxford University Press, Oxford. pp. 50–53.

Olson, Storrs L., and Helen F. James. 1982. Fossil birds from the Hawaiian Islands: evidence for wholesale extinction by man before western contact. *Science* 217:633–635.

Olson, Storrs L., and Helen F. James. 1984. The role of Polynesians in the extinction of the avifauna of the Hawaiian Islands. *In:* Paul S. Martin and Richard G. Klein (eds.), *Quaternary Extinctions: a Prehistoric Revolution.* University of Arizona Press, Tucson, AZ, pp. 768–780.

Ono, R. Dana, James D. Williams, and Anne Wagner. 1983. *Vanishing Fishes of North America.* Stone Wall Press, Inc., Washington, DC.

Opler, Paul A., and Herbert G. Baker, and Gordon W. Frankie. 1977. Recovery of tropical lowland forest ecosystems. *In:* J. Cairns, Jr., K.L. Dickson, and E.E. Herricks (eds.), *Recovery and Restoration of Damaged Ecosystems.* University Press of Virginia, Charlottesville, VA, pp. 379–421.

Orians, Gordon H., and William Kunin. Ecological uniqueness and loss of species. Proceedings of the Lake Wilderness Conference on the Conservation of Genetic Resources. University of Washington Press, Seattle, WA. In press.

OTA (Office of Technology Assessment, U.S. Congress). 1987. *Technologies to Maintain Biological Diversity.* OTA-F–330. U.S. Government Printing Office, Washington, DC.

Paine, Robert T. 1966. Food web complexity and species diversity. *American Naturalist* 100:65–75.

Paine, Robert T. 1980. Food webs: linkage, interaction strength and community infrastructure. *Journal of Animal Ecology* 49:667–685.

Palmberg, Christel. 1984. Genetic resources of arboreal fuelwood species for the improvement of rural living. *In:* Christopher W. Yeatman, David Kafton, and Garrison Wilkes (eds.), *Plant Genetic Resources: A Conservation Imperative.* AAAS Selected Symposium #87. Westview Press, Boulder, CO, pp. 223–239.

Phillips, Michael K., and Warren T. Parker. 1988. Red wolf recovery: a progress report. *Conservation Biology* 2:139–141.

Pielou, E.C. 1979. *Biogeography.* John Wiley & Sons, New York, NY.

Pimentel, David, E. Garnick, A. Berkowitz, S. Jacobson, S. Napolitano, P. Black, S. Valdes-Cogliano, B. Vinzant, E. Hudes, and S. Littman. 1980. Environmental quality and natural biota. *BioScience* 30:750–755.

Pimm, Stuart L. 1986. Community stability and structure. *In:* Michael E. Soulé (ed.), *Conservation Biology: The Science of Scarcity and Diversity.* Sinauer Associates Inc., Sunderland, MA, pp. 309–329.

Pimm, Stuart L., H. Lee Jones, and Jared Diamond. 1988. On the risk of extinction. *American Naturalist* 132:757–785.

Plucknett, Donald L., and Nigel J.H. Smith. 1986. Sustaining agricultural yields. *BioScience* 36:40–45.

Plucknett, Donald L., Nigel J.H. Smith, J.T. Williams, N. Murthi Anishetty. 1987. *Gene Banks and the World's Food.* Princeton University Press, Princeton, NJ.

Prance, Ghillean T. (ed.). 1982. *Biological Diversification in the Tropics.* Columbia University Press, New York, NY.

Prance, Ghillean T. Changes in the world flora. *In:* B.L. Turner II, W.C. Clark, R.W. Kates, J.F. Richards, J.T. Mathews, W.B. Meyer (eds.), *Earth as Transformed by Human Action.* Cambridge University Press, New York, NY. In press.

Prance, G[hillean] T., W. Balée, B.M. Boom, R.L. Carneiro. 1987. Quantitative ethnobotany and the case for conservation in Amazonia. *Conservation Biology* 1:296–310.

Prescott-Allen, Christine, and Robert Prescott-Allen. 1986. *The First Resource.* Yale University Press, New Haven, CT.

Prescott-Allen, Robert, and Christine Prescott-Allen. 1978. Sourcebook for a world conservation strategy: threatened vertebrates. General Assembly Paper GA.78/10 Addendum 6. International Union for Conservation of Nature and Natural Resources, Gland, Switzerland.

Prescott-Allen, Robert, and Christine Prescott-Allen. 1983. *Genes from the Wild.* International Institute for Environment and Development, London, U.K.

Prescott-Allen, Robert, and Christine Prescott-Allen. 1984. *In Situ Conservation of Wild Plant Genetic Resources: A Status Review and Action Plan.* Forest Resources Division, Forestry Department, Food and Agriculture Organization of the United Nations, Rome.

Principe, Peter P. The economic significance of plants and their constituents as drugs. *In: Economic and Medicinal Plant Research.* Vol. 3, Academic Press, New York, NY. pp. 1–17. In press.

Quinn, James F., and Susan P. Harrison. 1988. Effects of habitat fragmentation and isolation on species richness: evidence from biogeographic patterns. *Oecologia* 75:132–140.

Quinn, James F., and Alan Hastings. 1987. Extinction in subdivided habitats. *Conservation Biology* 1:198–208.

Rabinowitz, Deborah, Sara Cairns, and Theresa Dillon. 1986. Seven forms of rarity and their frequency in the flora of the British Isles. *In:* Michael E. Soulé (ed.), *Conservation Biology: The Science of Scarcity and Diversity*. Sinauer Associates, Sunderland, MA, pp. 182–204.

Räsänen, Matti E., Jukka S. Salo, Risto J. Kalliola. 1987. Fluvial perturbance in the western Amazon basin: regulation by long-term sub-Andean tectonics. *Science* 238:1398–1401.

Ralls, Katherine, and Jonathan Ballou. 1983. Extinction: lessons from zoos. *In:* Christine M. Schonewald-Cox, Steven M. Chambers, Bruce MacBryde, and W. Lawrence Thomas (eds.), *Genetics and Conservation*. Benjamin/Cummings Publishing Co., Menlo Park, CA, pp. 164–184.

Ransdell, Eric. 1989. Heavy artillery for horns of plenty. *U.S. News and World Report*, 20 February:61–64.

Rao, Y.S. 1988. Flash floods in southern Thailand. *Tiger Paper* 15(4):1–2.

Rapoport, E. 1982. *Areography*. Pergamon Press, Oxford, U.K.

Raup, David M. 1978. Cohort analysis of generic survivorship. *Paleobiology* 4:1–15.

Raup, David M. 1979. Size of the Permo-Triassic bottleneck and its evolutionary implications. *Science* 206:217–218.

Raup, David M., and J. John Sepkoski Jr. 1982. Mass extinctions in the marine fossil record. *Science* 215:1501–1503.

Raup, David M., and Steven M. Stanley. 1978. *Principles of Paleontology*. 2nd Edition, W.H. Freeman & Co., San Francisco, CA.

Raven, Peter H. 1976. Ethics and attitudes. *In:* J.B. Simmons, R.I. Beyer, P.E. Brandham, G.Ll. Lucas, and V.T.H. Parry (eds.), *Conservation of Threatened Plants*. Plenum Press, New York, NY, pp. 155–179.

Raven, Peter H. 1981. Research in botanical gardens. *Bot. Jahrb. Syst.* 102:53–72.

Raven, P.H. 1988a. Biological resources and global stability. *In:* S. Kawano, J.H. Connell, and T. Hidaka (eds.), *Evolution and Coadaptation in Biotic Communities*. University of Tokyo Press, Tokyo, pp. 3–27.

Raven, Peter H. 1988b. Our diminishing tropical forests. *In:* E.O. Wilson and Francis M. Peter (eds.), *Biodiversity*. National Academy Press, Washington, DC, pp. 119–122.

Raven, Peter H., and Daniel I. Axelrod. 1974. Angiosperm biogeography and past continental movements. *Annals of the Missouri Botanical Garden* 61:539–673.

Ray, G. Carleton. 1988. Ecological diversity in coastal zones and oceans. *In:* E.O. Wilson and Francis M. Peter (eds.), *Biodiversity*. National Academy Press, Washington, DC, pp. 36–50.

Ray, G. Carleton, Bruce P. Hayden, Arthur J. Bulger Jr., and M. Geraldine McCormick-Ray. Effects of global warming on marine biodiversity. *In:* R.L. Peters and T.E. Lovejoy (eds.), *Consequences of the Greenhouse Effect for Biological Diversity*. Yale University Press, New Haven, CT. In press.

Regal, Philip J. 1982. Pollination by wind and animals: ecology of geographic patterns. *Annual Review of Ecology and Systematics* 13:497–524.

Round, P.D. 1985. *The Status and Conservation of Resident Forest Birds in Thailand*. Association for the Conservation of Wildlife, Bangkok.

Ruggieri, George D. 1976. Drugs from the sea. *Science* 194:491–496.

Sader, Steven A., and Armond T. Joyce. 1988. Deforestation rates and trends in Costa Rica, 1940 to 1983. *Biotropica* 20:11–19.

Salati, Eneas, and Peter B. Vose. 1983. Depletion of tropical rain forests. *Ambio* 12:67–71.

Salm, Rodney V., and John R. Clark. 1984. *Marine and Coastal Protected Areas: A Guide for Planners and Managers.* International Union for the Conservation of Nature and Natural Resources, Gland, Switzerland.

Sayer, Jeffrey A., and Simon Stuart. 1988. Biological diversity and tropical forests. *Environmental Conservation* 15:193–194.

Schneider, Stephen H. 1989. The greenhouse effect: science and policy. *Science* 243:771–781.

Schwartzman, S[tephen], and M.H. Allegretti. 1987. *Extractive Production in the Amazon and the Rubber Tappers' Movement.* Environmental Defense Fund, Washington, DC.

Seal, Ulysses S. 1988. Intensive technology in the care of ex situ populations of vanishing species. *In:* E.O. Wilson and Francis M. Peter (eds.), *Biodiversity.* National Academy Press, Washington, DC, pp. 289–295.

Sedjo, Roger A. 1989. Managing genetic resources in sub-Saharan Africa: policy and institutional options. Report to the Africa Bureau, 24 February, U.S. Agency for International Development, Washington, DC.

Sepkoski, J. John Jr, and David M. Raup. 1986. Periodicity in marine extinction events. *In:* D.K. Elliott (ed.), *Dynamics of Extinction.* John Wiley and Sons, New York, NY, pp. 3–36.

Shaffer, Mark L. 1981. Minimum population sizes for species conservation. *BioScience* 31:131–134.

Sheeline, Lili. 1987. Is there a future in the wild for rhinos? *Traffic (U.S.A.)* 7(4):1–6.

Simberloff, Daniel. 1986a. Are we on the verge of a mass extinction in tropical rain forests? *In:* D.K. Elliott (ed.), *Dynamics of Extinction.* John Wiley & Sons, New York, NY, pp. 165–180.

Simberloff, D. 1986b. Introduced insects: a biogeographic and systematic perspective. *In:* H.A. Mooney and J.A. Drake (eds.), *Ecology of Biological Invasions of North America and Hawaii.* Ecological Studies 58. Springer-Verlag, New York, NY, pp. 3–26.

Simberloff, Daniel. 1987. The spotted owl fracas: mixing academic, applied, and political ecology. *Ecology* 68:766–772.

Simberloff, Daniel, and James Cox. 1987. Consequences and costs of conservation corridors. *Conservation Biology* 1:63–71.

Simberloff, Daniel, and Lawrence G. Abele. 1982. Refuge design and island biogeographic theory: effects of fragmentation. *American Naturalist* 120:41–50.

Simberloff, Daniel, and Nicholas Gotelli. 1984. Effects of insularisation on plant species richness in the prairie-forest ecotone. *Biological Conservation* 29:27–46.

Simpson, Beryl B., and Jürgen Haffer. 1978. Speciation patterns in the Amazonian forest biota. *Annual Review of Ecology and Systematics* 9:497–518.

Smith, Joel B., and Dennis A. Tirpak (eds.). 1988. *The Potential Effects of Global Climate Changes on the United States Vol. 2.* U.S. Environmental Protection Agency, Washington, DC.

Smith, S.V. 1978. Coral-reef area and contributions of reefs to processes and resources of the world's oceans. *Nature* 273:225–226.

Sommer, Adrian. 1976. Attempt at an assessment of the world's tropical moist forest. *Unasylva* 28(112 + 113):5–24.

Soulé, Michael E. 1980. Thresholds for survival: maintaining fitness and evolutionary potential. *In:* Michael E. Soulé and Bruce A. Wilcox (eds.), *Conservation Biology: An Evolutionary-Ecological Perspective.* Sinauer Associates, Sunderland, MA, pp. 151–169.

Soulé, Michael E. (ed.). 1987. *Viable Populations for Conservation.* Cambridge University Press, Cambridge, U.K.

Soulé, Michael E., and Daniel Simberloff. 1986. What do genetics and ecology tell us about the design of nature reserves? *Biological Conservation* 35:19–40.

Soulé, Michael E., and Kathryn A. Kohm (eds.). 1989. *Research Priorities for Conservation Biology.* Island Press, Covelo, CA.

Sousa, Wayne P. 1984. The role of disturbance in natural communities. *Annual Review of Ecology and Systematics* 15:353–391.

Speth, James Gustave. 1988. Environmental pollution. *In:* Harm J. De Blij (ed.), *Earth '88: Changing Geographic Perspectives.* National Geographic Society, Washington DC, pp. 262–283.

Stehli, Francis G., and John W. Wells. 1971. Diversity and age patterns in hermatypic corals. *Systematic Zoology* 20:115–125.

Stern, Klaus, and Laurence Roche. 1974. *Genetics of Forest Ecosystems.* Ecological Studies Vol. 6. Springer-Verlag, New York, NY.

Stout, Jean, and John Vandermeer. 1975. Comparison of species richness for stream-inhabiting insects in tropical and mid-latitude streams. *American Naturalist* 109:263–280.

Strong, Donald R., Jr. 1979. Biogeographic dynamics of insect-host plant communities. *Annual Review of Entomology* 24:89–119.

Tatum, L.A. 1971. The southern corn leaf blight epidemic. *Science* 171:1113–1116.

Teal, John M. 1962. Energy flow in the salt marsh ecosystem of Georgia. *Ecology* 43:614–624.

Terborgh, John. 1974. Preservation of natural diversity: the problem of extinction prone species. *BioScience* 24:715–722.

Terborgh, John. 1986a. Keystone plant resources in the tropical forest. *In:* Michael E. Soulé (ed.), *Conservation Biology: The Science of Scarcity and Diversity.* Sinauer Associates, Sunderland, MA, pp. 330–344.

Terborgh, John. 1986b. Community aspects of frugivory in tropical forests. *In:* A. Estrada and T.H. Fleming (eds.), *Frugivores and Seed Dispersal.* Dr. W. Junk Publishers, Dordrecht, pp. 371–384.

Terborgh, John, and Blair Winter. 1980. Some causes of extinction. *In:* Michael E. Soulé and Bruce A. Wilcox (eds.), *Conservation Biology: an Evolutionary-Ecological Perspective.* Sinauer Associates, Sunderland, MA, pp. 119–133.

Terborgh, John, and Blair Winter. 1983. A method for siting parks and reserves with special reference to Colombia and Ecuador. *Biological Conservation* 27:45–58.

Udvardy, Miklos D.F. 1975. A classification of the biogeographical provinces of the world. Occasional Paper 18. International Union for Conservation of Nature and Natural Resources, Gland, Switzerland.

Udvardy, Miklos D.F. 1984. A biogeographical classification system for terrestrial environments. *In:* Jeffrey A. McNeely and Kenton R. Miller (eds.), *National Parks, Conservation, and Development.* Smithsonian Institution Press, Washington, DC, pp. 34–38.

Uhl, Christopher, Carl Jordan, Kathleen Clark, Howard Clark, and Rafael Herrera. 1982.

Ecosystem recovery in Amazon caatinga forest after cutting, cutting and burning, and bulldozer clearing treatments. *Oikos* 38:313–320.

United Nations Population Fund. 1989. *The State of the World Population 1989.* United Nations Population Fund, New York, NY.

Valentine, James W., Theodore C. Foin, and David Peart. 1978. A provincial model of Phanerozoic marine diversity. *Paleobiology* 4:55–66.

Vatikiotis, Michael. 1989. Whither the vine? *Far Eastern Economic Review* 13 April:50.

Vermeij, Geerat J. 1978. *Biogeography and Adaptation: Patterns of Marine Life.* Harvard University Press, Cambridge, MA.

Vernhes, Jane Robertson. 1989. Biosphere reserves: the beginnings, the present, and the future challenges. *In:* William P. Gregg, Jr., Stanley L. Krugman, and James D. Wood, Jr. (eds.), Proceedings of the Symposium on Biosphere Reserves, September 11–18, 1987, Estes Park, CO. U.S. National Park Service, Atlanta, GA, pp. 7–20.

Vernon, J.E.N. 1986. *Corals of Australia and the Indo-Pacific.* Angus and Robertson, London U.K.

Vietmeyer, Noel D. 1986. Lesser-known plants of potential use in agriculture and forestry. *Science* 232:1379–1384.

Vitousek, P[eter] M. 1986. Biological invasions and ecosystem properties: can species make a difference? *In:* H.A. Mooney and J.A. Drake (eds.), *Ecology of Biological Invasions of North America and Hawaii.* Ecological Studies 58. Springer-Verlag, New York, NY, pp. 163–176.

Vitousek, Peter M. 1988. Diversity and biological invasions of oceanic islands. *In:* E.O. Wilson and Francis M. Peter (eds.), *Biodiversity.* National Academy Press, Washington, DC, pp. 181–189.

Vitousek, Peter M., Paul R. Ehrlich, Anne H. Ehrlich, and Pamela A. Matson. 1986. Human appropriation of the products of photosynthesis. *BioScience* 36:368–373.

von Fürer-Haimendorf, Christoph. 1964. *The Sherpas of Nepal.* University of California Press, Berkeley, CA.

Wagner, Warren L., Derral R. Herbst, and S.H. Sohmer. *Manual of the Flowering Plants of Hawaii.* University of Hawaii Press and Bishop Museum Press, Honolulu, HI. In press.

WEC (World Energy Conference). 1987. *Energy: Needs and Expectations.* Proceedings of the 13th Congress of the World Energy Conference, Cannes, October 5–11th, 1986. WEC, London.

White, F. 1983. *The Vegetation of Africa: A Descriptive Memoir to Accompany the Unesco/AETFAT/UNSO Vegetation Map of Africa.* UNESCO, Paris.

Whittaker, Robert H. 1975. *Communities and Ecosystems* (2nd ed.). MacMillan Publishing Co., Inc., New York, NY.

Wilcove, David S. 1988. *National Forests: Policies for the Future, Vol. 2: Protecting Biological Diversity.* The Wilderness Society, Washington, DC.

Wilcove, David S., and Robert M. May. 1986. National park boundaries and ecological realities. *Nature* 324:206–207.

Wildt, David E., and Ulysses S. Seal (eds.). 1988. Research priorities for single species conservation biology. Proceedings of a Workshop at the National Zoological Park, Washington, DC, 13–16 November 1988.

Wilkes, H. Garrison. 1972. Maize and its wild relatives. *Science* 177:1071–1077.

Wilkes, H. Garrison. 1979. Mexico and Central America as a centre for the origin of agriculture and the evolution of maize. *Crop Improvement* 6:1–18.

Wilkes, Garrison. 1983. Current status of crop plant germplasm. *CRC Critical Reviews in Plant Science* 1(2):133–181.

Wilkes, Garrison. 1985. Germplasm conservation toward the year 2000: potential for new crops and enhancement of present crops. *In:* Christopher W. Yeatman, David Kafton, and Garrison Wilkes (eds.), *Plant Genetic Resources: A Conservation Imperative.* American Association for the Advancement of Science Selected Symposium #87. Westview Press, Boulder, CO, pp. 131–164.

Wilkes, Garrison. (a). Teosinte in Mexico as a model for in situ conservation: the challenge. *Maydica.* In press.

Wilkes, Garrison. (b). Germplasm preservation: objectives and needs. Proceedings of the Beltsville Symposium XIII Biotic Diversity and Germplasm Preservation: Global Imperative. U.S. Department of Agriculture, Washington, DC. In press.

Williams, Jack E., David B. Bowman, James E. Brooks, Anthony A. Echelle, Robert J. Edwards, Dean A. Hendrickson, and Jerry J. Landye. 1985. Endangered aquatic ecosystems in North American deserts with a list of vanishing fishes of the region. *Journal of the Arizona-Nevada Academy of Science* 20(1):1–61.

Williams, J.T. 1984. A decade of crop genetic resources research. *In:* J.H.W. Holden and J.T. Williams (eds.), *Crop Genetic Resources: Conservation and Evaluation.* Allen and Unwin, London, U.K., pp. 1–17.

Wilson, E[dward] O. 1988a. The current state of biological diversity. *In:* E.O. Wilson and Francis M. Peter (eds.), *Biodiversity.* National Academy Press, Washington, DC, pp. 3–18.

Wilson, E[dward] O. 1988b. The diversity of life. *In:* Harm J. De Blij (ed.), *Earth '88: Changing Geographic Perspectives.* National Geographic Society, Washington, DC, pp. 68–78.

Wong, Marina. 1985. Understory birds as indicators of regeneration in a patch of selectively logged west Malaysian rainforest. *In:* A.W. Diamond and T.E. Lovejoy (eds.), *Conservation of Tropical Forest Birds.* Technical Publication No. 4. International Council for Bird Preservation, Cambridge, U.K., pp. 249–263.

Wood, D. 1988. Introduced crops in developing countries. *Food Policy* 13(2):167–177.

World Bank. 1988. *World Development Report 1988.* World Bank, Washington, DC.

WRI (World Resources Institute). 1989. *Natural Endowments: Financing Resource Conservation for Development.* WRI, Washington, DC.

WRI/IIED (WRI/International Institute for Environment and Development). 1986. *World Resources 1986.* Basic Books, NY.

WRI/IIED. 1987. *World Resources 1987.* Basic Books, NY.

WRI/IIED. 1988. *World Resources 1988–1989.* Basic Books, NY.

Wyler, David J. 1983. Malaria—resurgence, resistance, and research. *New England Journal of Medicine* 308:875-878.

Zaret, Thomas M., and R.T. Paine. 1973. Species introduction in a tropical lake. *Science* 182:449-455.

World Resources Institute

1709 New York Avenue, N.W.
Washington, D.C. 20006, U.S.A.

WRI's Board of Directors:
Matthew Nimetz
Chairman
Roger W. Sant
Vice Chairman
John H. Adams
Kazuo Aichi
Robert O. Anderson
Manuel Arango
Robert O. Blake
John E. Bryson
John E. Cantlon
Pamela G. Carlton
Ward B. Chamberlin
Edwin C. Cohen
Louisa C. Duemling
Alice F. Emerson
John Firor
Shinji Fukukawa
Cynthia R. Helms
Curtis A. Hessler
Martin Holdgate
Maria Tereza Jorge Padua
Jonathan Lash
Thomas E. Lovejoy
C. Payne Lucas
Alan R. McFarland, Jr.
Robert S. McNamara
Scott McVay
Paulo Nogueira-Neto
Saburo Okita
Ronald L. Olson
Ruth Patrick
Alfred M. Rankin, Jr.
Stephan Schmidheiny
Bruce Smart
James Gustave Speth
Maurice Strong
M.S. Swaminathan
Mostafa K. Tolba
Russell E. Train
Alvaro Umaña
Victor L. Urquidi
George M. Woodwell

Jonathan Lash
President
J. Alan Brewster
Senior Vice President
Walter V. Reid
Vice President for Program
Donna W. Wise
Vice President for Policy Affairs
Jessica T. Mathews
Vice President
Robert Repetto
Vice President and Senior Economist
Wallace D. Bowman
Secretary-Treasurer

The World Resources Institute (WRI) is a policy research center created in late 1982 to help governments, international organizations, and private business address a fundamental question: How can societies meet basic human needs and nurture economic growth without undermining the natural resources and environmental integrity on which life, economic vitality, and international security depend?

Two dominant concerns influence WRI's choice of projects and other activities:

The destructive effects of poor resource management on economic development and the alleviation of poverty in developing countries; and

The new generation of globally important environmental and resource problems that threaten the economic and environmental interests of the United States and other industrial countries and that have not been addressed with authority in their laws.

The Institute's current areas of policy research include tropical forests, biological diversity, sustainable agriculture, energy, climate change, atmospheric pollution, economic incentives for sustainable development, and resource and environmental information.

WRI's research is aimed at providing accurate information about global resources and population, identifying emerging issues, and developing politically and economically workable proposals.

In developing countries, WRI provides field services and technical program support for governments and non-governmental organizations trying to manage natural resources sustainably.

WRI's work is carried out by an interdisciplinary staff of scientists and experts augmented by a network of formal advisors, collaborators, and cooperating institutions in 50 countries.

WRI is funded by private foundations, United Nations and governmental agencies, corporations, and concerned individuals.

0936

0996